DIGITAL FILTERS
Third Edition

PRENTICE HALL SIGNAL PROCESSING SERIES

Alan V. Oppenheim, Editor

AT&T

DIGITAL FILTERS
Third Edition

R. W. HAMMING

Naval Postgraduate School
Monterey, California

PRENTICE HALL, ENGLEWOOD CLIFFS, NEW JERSEY 07632

LIBRARY OF CONGRESS
Library of Congress Cataloging-in-Publication Data

Hamming, R. W. (Richard Wesley)
    Digital filters / R.W. Hamming. -- 3rd ed.
        p.   cm. -- (Prentice Hall signal processing series)
    Bibliography: p.
    Includes index.
    ISBN 0-13-212812-8
    1. Digital filters (Mathematics)   I. Title. II. Series.
QA297.H26 1983
    621.38'043--dc19   88-19598
                                                               CIP

Editorial/production supervision: Merrill Peterson
Cover design: Karen Stephens
Manufacturing buyer: Mary Noonan

© 1989 by Prentice-Hall, Inc.
A Division of Simon & Schuster
Englewood Cliffs, New Jersey 07632

Printed in the United States of America

10 9 8 7 6 5 4 3 2 1

ISBN 0-13-212812-8

Prentice-Hall International (UK) Limited, *London*
Prentice-Hall of Australia Pty. Limited, *Sydney*
Prentice-Hall Canada Inc., *Toronto*
Prentice-Hall Hispanoamericana, S.A., *Mexico*
Prentice-Hall of India Private Limited, *New Delhi*
Prentice-Hall of Japan, Inc., *Tokyo*
Simon and Schuster Asia Pte. Ltd., *Singapore*
Editora Prentice-Hall do Brasil, Ltda., *Rio de Janeiro*

Contents

Preface
to the Third Edition

This edition retains the same basic approach of the earlier editions of stressing fundamentals; however, some changes have been made to reflect the fact that increasingly often a digital filter course is the first course in electrical engineering and the field of signal processing. To meet these needs two main changes have been made: (1) the inclusion of more material on the z-transform, which is often used in later courses (though the constant use of the formalism tends to obscure the ideas behind the manipulations), and (2) the inclusion of more examples and exercises. There are, of course, many minor changes to clarify and adapt the material to current uses.

In the years since I wrote the first edition I have become increasingly convinced of the need for a very elementary treatment of the subject of digital filters. The need for an elementary introduction comes from the fact that many of the people who most need the knowledge are not mathematically sophisticated and do not have an elaborate electrical engineering background. Thus this book assumes *only* a knowledge of the calculus and a smattering of statistics (which is reviewed in the text). It does not assume any electrical engineering background knowledge. Actually, experience seems to show that a prior knowledge of the corresponding theory of analog filters often causes more harm than good! Digital filtering is not simply converting from analog to digital filters; it is a fundamentally different way of thinking about the topic of signal processing, and many of the ideas and limitations of the analog method have no counterpart in the digital form.

The subject of digital filters is the natural introduction to the broad, fundamental field of signal processing. The power and basic simplicity of digital signal processing over the older analog is so great that whenever possible we are converting present analog systems to an equivalent digital form. But much more important, digital signaling allows fundamentally new things to be done easily. The availability of modern integrated-circuit chips, as well as micro- and minicomputers, has greatly expanded the application of digital filters.

Digital signals occur in many places. The telephone company is rapidly converting to the use of digital signals to represent the human voice. Even radio, television, and hi-fi sound systems are moving toward the all digital methods since they provide such superior fidelity and freedom from noise, as well as much more flexible signal processing. The space shots use digital signaling to transmit the information from the planets back to Earth, including the extremely detailed pictures (which were often processed digitally here on Earth to extract further information and to form alternate views of what was originally captured by the cameras in space). Most records of laboratory experiments are now recorded in digital form, from isolated measurements using a digital voltmeter to the automatic recording of entire sets of functions via a digital computer. Thus these signals are immediately ready for digital signal processing to extract the message that the experiment was designed to reveal. Economic data, from stock market prices and averages to the Cost of Living Index of the Bureau of Labor Statistics, occur only in digital form.

Digital filtering includes the processes of smoothing, predicting, differentiating, integrating, separation of signals, and removal of noise from a signal. Thus many people who do such things are actually using digital filters without realizing that they are; being unacquainted with the theory, they neither understand what they have done nor the possibilities of what they might have done. Computer people very often find themselves involved in filtering signals when they have had no appropriate training at all. Their needs are especially catered to in this book.

Because the same ideas arise in many fields there are many cross connections between the fields that can be exploited. Unfortunately each field seems to go its own way (while reinventing the wheel) and to develop its own jargon for exactly the same ideas that are used elsewhere. One goal of this revision is to expose and reduce this elaborate jargon equivalence from the various fields of application and to provide a unified approach to the whole field. We will adopt the simplest, most easily understood words to describe what is going on and exhibit lists of the equivalent words from related fields. We will also use only the simplest, most direct mathematical tools and shun fancy mathematics whenever possible.

This book concentrates on linear signal processing; the main exceptions are the examination of roundoff effects and a brief mention of Kalman filters, which adapt themselves to the signal they are receiving.

The fundamental tool of digital filtering is the frequency approach, which is based on the use of sines and cosines rather than on the use of polynomials (as is conventional in many fields such as numerical analysis and much of statistics). The frequency approach, which leads to the spectrum, has been the principal method of opening the black boxes of nature. Examples run from the early study of the structure of the atom (using spectral lines as the observations) through quantum mechanics (which arose from the study of the spectrum of black-body radiation) to the modern methods of studying a system (for purposes of modeling and control) via the spectrum of the output as it is related to the input.

There appears to be a deep emotional resistance to the frequency approach. And even electrical engineers who use it daily often have only a slight understanding of *why* they are using the eigenfunction approach and the role of the eigenvalues. In numerical analysis there is almost complete antipathy to the frequency approach, while in statistics there is a great fondness for polynomials (without ever examining the question of which set of functions is appropriate). This book shows clearly why the sines and cosines are the natural, the proper, the characteristic functions to use in many situations. It also approaches cautiously the usual traumatic experience (for most people) of going from the real sines and cosines to the complex exponentials with the mysterious $\sqrt{-1}$; their greater convenience in use eventually compensates for the initial troubles and provides more insight.

The text includes an accurate (but not excessively rigorous) introduction to the necessary mathematics. In each case the formal mathematics is postponed until the need for it is clearly seen. We are interested in presenting the *ideas* of the field and will generally not give the "best" methods for designing very complex filters; in an elementary course it is proper to give elementary, broadly applicable design methods, and then show how these can be refined to meet a very wide range of design criteria. Because it is an elementary text, references to advanced papers and books are of little use to the reader. Instead we refer to a few standard texts where more advanced material and references can be found. The references to these books are indicated in the text by [L,p], where L is the book label given at the end of this book, and p is the page(s) where it can be found. References [IEEE-1 and 2] give a complete bibliography for most topics that arise.

There is a deliberate repetition in the presentation of the material. Experience shows that the learner often becomes so involved in the immediate details of designing a filter that where and how the topic fits into the whole plan is lost. Furthermore, confusion often arises when the same ideas and mathematical tools are used in seemingly very different situations. It is also true that filters are designed to process data, but experience shows that the display of large sets of data that have been processed communicates very little to the beginner. Thus such plots are seldom given, even though the learner needs to be reminded that the ultimate test of a filter is how well it processes a signal, not how elegant the derivation is.

As always an author is deeply indebted to others, in this case to his many colleagues at Bell Laboratories. Special mention should go to Professor J. W. Tukey (of Princeton University) and to J. F. Kaiser, who first taught him most of what is presented here. Thanks are also due to Roger Pinkham and the many students of the short courses who used the first two editions; their questions and reactions have been important in many places of this revision. They have also strengthened the author's belief in the basic rightness of giving as simple an approach as possible and of keeping rigorous mathematics in its proper place. Finally, thanks are due to the Naval Post-graduate School for providing an atmosphere suitable for thinking deeply about the problems of teaching.

R. W. Hamming

DIGITAL FILTERS
Third Edition

1

Introduction

1.1 WHAT IS A DIGITAL FILTER?

In our current technical society we often measure a continuously varying quantity. Some examples include blood pressure, earthquake displacements, voltage from a voice signal in a telephone conversation, brightness of a star, population of a city, waves falling on a beach, and the probability of death. All these measurements vary with time; we regard them as functions of time: $u(t)$ in mathematical notation. And we may be concerned with blood pressure measurements from moment to moment or from year to year. Furthermore, we may be concerned with functions whose independent variable is not time, for example the number of particles that decay in a physics experiment as a function of the energy of the emitted particle. Usually these variables can be regarded as varying continuously (analog signals) even if, as with the population of a city, a bacterial colony, or the number of particles in the physics experiment, the number being measured must change by unit amounts.

For technical reasons, instead of the signal $u(t)$, we usually record *equally spaced samples* u_n of the function $u(t)$. The famous *sampling theorem*, which will be discussed in Chapter 8, gives the conditions on the signal that justify this sampling process. Moreover, when the samples are taken they are not recorded with infinite precision but are rounded off (sometimes chopped off) to comparatively few digits (see Figure 1.1-1). This procedure is often called *quantizing* the samples. It is these quantized samples that are available for the processing that we do. We do the processing in order to understand what the function samples u_n reveal about the underlying phenomena that gave rise to the observations, and digital filters are the main processing tool.

1

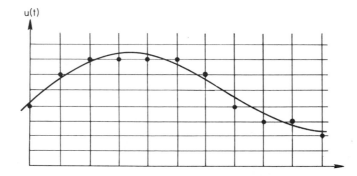

FIGURE 1.1-1 SAMPLING AND QUANTIZATION OF A SIGNAL

It is necessary to emphasize that the samples are *assumed* to be equally spaced; any error or *noise* is in the measurements u_n. Fortunately, this assumption is approximately true in most applications.

Suppose that the sequence of numbers $\{u_n\}$ is such a set of equally spaced measurements of some quantity $u(t)$, where n is an integer and t is a continuous variable. Typically, t represents time, but not necessarily so. We are using the notation $u_n = u(n)$. The simplest kinds of filters are the *nonrecursive filters*; they are defined by the linear formula

$$y_n = \sum_{k=-\infty}^{\infty} c_k u_{n-k} \qquad (1.1\text{-}1)$$

The coefficients c_k are the constants of the filter, the u_{n-k} are the input data, and the y_n are the outputs. Figure 1.1-2 shows how this formula is computed. Imagine two strips of paper. On the first strip, written one below the other, are the data values u_{n-k}. On the second strip, with the values written in the *reverse direction* (from bottom to top), are the filter coefficients c_k. The zero subscript of one is opposite the n subscript value of the other (either way). The output y_n is the sum of all the products $c_k u_{n-k}$. Having computed one value, one strip, say the coefficient strip, is moved one space down, and the new set of products is computed to give the new output y_{n+1}. Each output is the result of adding all the products formed from the proper displacement between the two zero-subscripted terms. In the computer, of course, it is the data that is "run past" the coefficient array $\{c_k\}$.

This process is basic and is called a *convolution* of the data with the coefficients. It does not matter which strip is written in the reverse order; the result is the same. So the convolution of u_n with the coefficients c_k is the same as the convolution of the coefficients c_k with the data u_n.

In practice, the number of products we can handle must be finite. It is usual to assume that the length of the run of nonzero coefficients c_k is much shorter than is the run of data y_n. Once in a while it is useful to regard the

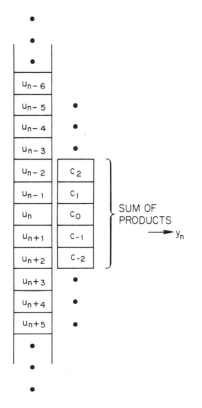

FIGURE 1.1-2 A NONRECURSIVE DIGITAL FILTER

c_k coefficients as part of an infinite array with many zero coefficients, but it is usually preferable to think of the array $\{c_k\}$ as being finite and to ignore the zero terms beyond the end of the array. Equation (1.1-1) becomes, therefore,

$$y_n = \sum_{k=-N}^{N} c_k u_{n-k} \tag{1.1-2}$$

Thus the second strip (of coefficients c_k) in Figure 1.1-2 is comparatively shorter than is the first strip (of data u_n).

Various special cases of this formula occur frequently and should be familiar to most readers. Indeed, such formulas are so commonplace that a book could be devoted to their listing. In the case of five nonzero coefficients c_k, where all the coefficients that are not zero have the same value, we have the familiar smoothing by 5s formula (derived in Section 3.2)

$$y_n = \tfrac{1}{5}(u_{n-2} + u_{n-1} + u_n + u_{n+1} + u_{n+2}) \tag{1.1-3}$$

Another example is the least-squares smoothing formula derived by passing a least-squares cubic through five equally spaced values u_n and using the value of the cubic at the midpoint as the smoothed value. The formula for this smoothed value (which will be derived in Section 3.3) is

$$y_n = \tfrac{1}{35}(-3u_{n-2} + 12u_{n-1} + 17u_n + 12u_{n+1} - 3u_{n+2}) \quad (1.1\text{-}4)$$

Many other formulas, such as those for predicting stock market prices, as well as other time series, also are nonrecursive filters.

Nonrecursive filters occur in many different fields and, as a result, have acquired many different names. Among the disguises are the following:

Finite impulse response filter

FIR filter

Transversal filter

Tapped delay line filter

Moving average filter

We shall use the name *nonrecursive* as it is the simplest to understand from its name, and it contrasts with the name *recursive filter*, which we will soon introduce.

The concept of a *window* is perhaps the most confusing concept in the whole subject, so we now introduce it in these simple cases. We can think of the preceding formulas as if we were looking at the data u_{n-k} through a *window of coefficients* c_k (see Figure 1.1-3). As we slide the strip of coefficients along the data, we see the data in the form of the output y_n, which

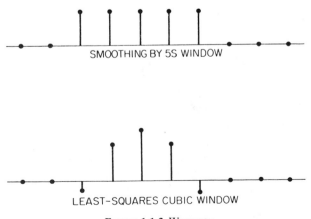

SMOOTHING BY 5S WINDOW

LEAST-SQUARES CUBIC WINDOW

Figure 1.1-3 Windows

is the running weighted average of the original data u_n. It is as if we saw the data through a translucent (not transparent) window where the window was tinted according to the coefficients c_k. In the smoothing by $5s$, all data values get through the translucent window with the same amount, $\frac{1}{5}$; in the second example they come through the window with varying weights. (Don't let any negative weights bother you, since we are merely using a manner of speaking when we use the words "translucent window.")

When we use not only data values to compute the output values y_n but also use other values of the output, we have a formula of the form

$$y_n = \sum_{-\infty}^{\infty} c_k u_{n-k} + \sum_{-\infty}^{\infty} d_k y_{n-k}$$

where both the c_k and the d_k are constants. In this case it is usual to limit the range of nonzero coefficients to current and past values of the data u_n and to only past values of the output y_n. Furthermore, again the number of products that can be computed in practice must be finite. Thus the formula is usually written in the form

$$y_n = \sum_{0}^{N} c_k u_{n-k} + \sum_{1}^{M} d_k y_{n-k} \qquad (1.1\text{-}5)$$

where there may be some zero coefficients. These are called *recursive* filters (see Figure 1.1-4). Some equivalent names follow:

Infinite impulse response filter

IIR filter

Ladder filter

Lattice filter

Wave digital filter

Autoregressive moving average filter

ARMA filter

Autoregressive integrated moving average filter

ARIMA filter

We shall use the name *recursive* filter. A recursive digital filter is simply a linear difference equation with constant coefficients and nothing more; in practice it may be realized by a short program on a general purpose digital computer or by a special purpose integrated circuit chip.

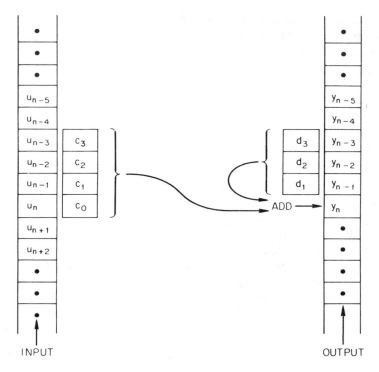

INPUT   OUTPUT

FIGURE 1.1-4 RECURSIVE DIGITAL FILTER

A familiar example (from the calculus) of a recursive filter is the trapezoid rule for integration

$$y_n = y_{n-1} + \tfrac{1}{2}[u_n + u_{n-1}] \tag{1.1-6}$$

It is immediately obvious that a recursive filter can, as it were, remember all the past data, since the y_{n-1} value on the right side of the equation enters into the computation of the new value y_n, and hence into the computation of y_{n+1}, y_{n+2}, and so on. In this way the initial condition for the integration is "remembered" throughout the entire estimation of the integral.

Other examples of a recursive digital filter are the exponential smoothing forecast

$$y_{n+1} = au_{n+1} + (1 - a)y_n \qquad (0 < a < 1)$$

and the trend indicator

$$T_n = c[u_n - u_{n-1}] + (1 - c)T_{n-1} \qquad (0 < c < 1)$$

As is customary, we have set aside recursive filters that use *future* values, values beyond the currently computed value. If we used future values beyond the current y_n, we would have to solve a system of linear algebraic equations, and this is a lot of computing. At times it is worth it, but often we have only past computed values and the current value of the data. Filters that use only past and current values of the data are called *causal*, for if time is the independent variable, they do not react to future events but only past ones (causes).

It is worth a careful note, however, that more and more often all the data of an experiment is recorded on a magnetic tape or other storage medium *before* any data processing is done. In such cases the restriction to causal filters is plainly foolish. Future values are available! There are, of course, many situations in which the data must be reduced and used as they come in, and in such cases the restriction to causal filters is natural.

The student may wonder where we get the starting values of the y_n. Once well going they, of course, come from previous computations, but how to start? The custom of assuming that the missing values y are to be taken as zeros is very dubious. This assumption usually amounts to putting a sharp discontinuity into the function y_n, and since as noted previously the recursive filter remembers the past, it follows that these zero values continue to affect the computation for some time, if not indefinitely. It is evident in the simple example of the trapezoid integration that the needed starting value of y is the starting area, usually taken to be zero, but not necessarily so.

We have said it before, but it is necessary to say again that the coefficients c_k and d_k of the filter are assumed to be constants. Such filters are called *time-invariant filters* and are the filters most used in practice. Time-varying filters are occasionally useful and will be briefly touched upon in this text.

Finally, it should be realized that in practice all computing must be done with finite-length numbers. The process of quantization affects not only the input numbers, but it may affect all the internal (to the filter) arithmetic that is done. Consequently, there are roundoff errors in the final output numbers y_n. It is often convenient to *think* in terms of infinite precision arithmetic and perfect input data u_n; but in the end we must deal with reality. Furthermore, the details of the way we arrange to do the arithmetic can affect the accuracy of the output numbers. We will look at this topic more closely in the closing chapters.

1.2 WHY SHOULD WE CARE ABOUT DIGITAL FILTERS?

The word *filter* is derived from electrical engineering, where filters are used to transform electrical signals from one form to another, especially to eliminate (filter out) various frequencies in a signal. As we have already seen, a digital filter is a linear combination of the input data u_n and possibly the output data y_n and includes many of the operations that we do when processing a signal.

For convenience we suppose that we sample at unit time and that we represent the nth sample of the signal as u_n. Thus we may think of blood pressure, a brain wave, the height of a wave on a beach, or a stock market price as a continuous signal that we have sampled (and quantized) at unit times in order to obtain our sequence of data u_n. In the stock market case we often take the integral over a period, say a week, and record only the total amount of stock sold each week, although the underlying idea of continuous variation of the price or whatever else is being measured (say the rate at which shares are traded) still exists. Given such a signal, we may want to differentiate, integrate, sum, difference, smooth, extrapolate, analyze for periodicity, or possibly remove the noise; all these, and many others, are linear operations. Therefore, in the digital form, the operations are *digital filters*.

Widespread use of mini- and microcomputers in science, medicine, and engineering has greatly increased the number of digital signals recorded and processed. Since we are already processing such data in a linear fashion, it is necessary to understand the alterations and distortions that these filters produce. Moreover, because digital transmission is so much more noise resistant than is analog signal transmission, a world dominated by digital transmission is rapidly approaching. Thus again we are impelled to study exactly what digital filters do, or can be designed to do, to various signals.

Applications of digital filters now greatly transcend those that arise in electrical engineering. As a result, it is necessary to redefine and remove some of the restrictions that were natural to electrical engineering at the time when digital filters were emerging from the classical electrical analog filters. The student should carefully note that we are sometimes making *different definitions than those that frequently occur in older electrical engineering texts*. It is necessary to do this because we have a larger view of the field of applications; we include applications to numerical analysis and statistics, for example, as well as to other fields.

Occasionally you read about *sampled data systems*. Here the signal is sampled, but the sampled value is *not* quantized. Such systems will not be considered in this book.

We will always assume that the samples u_n are unit spaced and begin at $t_0 = 0$. If they are not, it is easy to find the linear transformation that will make them so. Let the original data be at

$$t_n = t_0 + n \, \Delta t \qquad (n = 0, 1, 2, \ldots)$$

To find the transformation, we simply assume the form

$$t'_n = a t_n + b$$

and impose the two conditions

$$t_0' = 0 = at_0 + b$$
$$t_1' = 1 = a(t_0 + \Delta t) + b$$

Solve (by subtraction)

$$1 = a\,\Delta t$$
$$b = -at_0$$

Therefore,

$$t_n' = \frac{1}{\Delta t}(t_n - t_0) \qquad (1.2\text{-}1)$$

It is easy to see that this is the desired transformation.

The method of derivation should be learned rather than the result memorized.

Exercises

1.2-1 Find the standard transformation for the data: $t_0 = 10$, $t_1 = 12$, $t_2 = 14$, . . . , *Answer:* $t' = \frac{1}{2}(t_n - 10)$.

1.2-2 Find the transformation that moves the sample points $t_0 = 0.100$, $t_1 = 0.112$, $t_2 = 0.124$, . . . , to the standard form.

1.2-3 List ten sources, other than those in the text, of signals that might be filtered.

1.2-4 Write Simpson's integration formula as a recursive filter.

1.2-5 Compute the first five output values of the filter

$$y_n = ay_{n-1} + u_n$$

for the input $y_0 = 0$, $u_n = 1$ for all n.

1.2-6 Compute the successive values of the filter

$$y_n = ay_{n-1} + u_n$$

where $y_0 = 0$, $u_1 = 1$, and all other $u_n = 0$. Give the formula for y_n.

1.2-7 Compact disc (CD) recordings use 44.1 kH (kH = kiloHertz = 1000 times a second). Find the appropriate transformation.

1.3 HOW SHALL WE TREAT THE SUBJECT?

Much of the theory, both as to the design and use of digital filters, originated in the field of analog filters. If one is already familiar with the field, then it might be reasonable to build on this knowledge. Today, however, the average person who needs to know about digital filters has no such background, and so it is foolish to base the development on the analog approach. Consequently, we assume no such familiarity and will only mention the corresponding jargon when necessary.

The statistics field has also contributed extensively to the theory of digital filters. In particular, the subject of time series is closely related and has contributed its own elaborate and confusing jargon.

Textbooks in numerical analysis have many formulas that are linear combinations of equally spaced data, and thus such formulas are equivalent to digital filters. Since the elements of numerical analysis are now more widely known than those of other fields of application, we will select many of our examples from numerical analysis. Furthermore it is very often needed in practice.

The fundamental approach common to all the special fields is based on (1) the Fourier series, both discrete and continuous, and (2) the use of the Fourier integral. They are the mathematical tools for understanding and manipulating linear formulas, and we must take the time to develop them, for they are rarely taught outside of electrical engineering courses these days. However, we will avoid becoming too involved with mathematical rigor, which all too often tends to become rigor mortis. Nor do we develop all the mathematical theory before showing its use; instead, we regularly give applications of the theory just covered in order to show both its relevance and its use. In this way, we hope that much of the mathematics will become more obvious to the nonmathematically inclined.

1.4 GENERAL-PURPOSE VERSUS SPECIAL-PURPOSE COMPUTERS

Digital filtering is done using both special- and general-purpose digital computers. Even though numerical computations also use both types, most introductory textbooks on the subject deal only with computing done on general-purpose computers; similarly, most of the discussion in this book is confined to filtering done on general-purpose computers.

This remark should not be interpreted as meaning that the field of special-purpose computers is unimportant. Rather it is an indication that computation on a general-purpose computer is usually much less restrictive. Therefore, in a first presentation, we concentrate on the main ideas, while ignoring the details of the particular computer being used. Special-purpose digital computers are rapidly increasing in importance, primarily because of the availability of inexpensive, large-scale integrated circuits, as well as the fact that many of the operations that we wish to perform are sometimes beyond the scope (in either an economic or a time sense, or both) of current (and foreseeable) general-purpose computers.

1.5 ASSUMED STATISTICAL BACKGROUND

We need to take a brief look at statistics. A set of measurements is called a *sample*. The word "sample" is used both for an individual measurement and for the set of measurements, even if they are repeated measurements of the same thing. This usage occurs because the statistician is thinking of an underlying *population* or *ensemble* of possible measurements and you have obtained one possible set of results (one realization). The statistician is concerned with the probability of obtaining the particular observed result and with the effects of repetitions of the experiment. The measurements of a sample may all be at a single point. For instance, the sample can be a number of measurements of the length of a particular wire. The measurements can also be scattered at various places in the range of a function, for example, the velocity of a boat at various times of a day.

Often a *model* for the distribution of the measurements must be found; we want to think about the ensemble from which we have drawn the particular sample.

To illustrate, if L is the measured length of the wire just mentioned, then we model these measurements by

$$P\{L \leq x\} = P(x)$$

For $P\{L \leq x\}$ read "probability that L is less than or equal to x." $P(x)$ is thus the probability that the measured length L is less than or equal to x; $P(x)$ is called the *cumulative distribution function for L*. In many situations $P(x)$ has a derivative $p(x)$; that is,

$$\frac{dP}{dx} = p(x) \quad \text{and} \quad P\{a < L \leq b\} = \int_a^b p(x)\, dx \qquad (1.5\text{-}1)$$

Then $p(x)$ is called either the *density* or the *probability density for L*.

A common density that occurs in such situations as measuring the length of a piece of wire is the gaussian (or normal) distribution

$$p(x) = \frac{1}{\sigma\sqrt{2\pi}}\, e^{-(x-\mu)^2/2\sigma^2} \qquad (-\infty < x < \infty) \qquad (1.5\text{-}2)$$

where μ and σ are parameters whose values depend on the particular situation being modeled. See Figure 1.5-1(b).

Another example of a model occurs in roundoff theory. It is reasonable to suppose that the roundoff error made when a number is quantized (rounded off) is "uniformly distributed" from $-\frac{1}{2}$ to $\frac{1}{2}$ in the last digit kept. Therefore,

$$p(x) \equiv \begin{cases} 1, & -\frac{1}{2} \le x \le \frac{1}{2} \\ 0, & |x| > \frac{1}{2} \end{cases} \qquad (1.5\text{-}3)$$

See Figure 1.5-1(a).

A commonly computed characteristic of a random quantity such as L or, alternatively, of a density $p(x)$ is the *average* or *expected value* (also called *mean value*). It is denoted Ave $\{L\}$ or $E\{L\}$ and is defined by

$$\text{Ave } \{L\} \equiv E\{L\} \equiv \int_{-\infty}^{\infty} x p(x) \, dx \qquad (1.5\text{-}4)$$

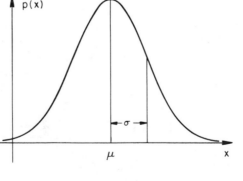

(a) ROUNDOFF DISTRIBUTION

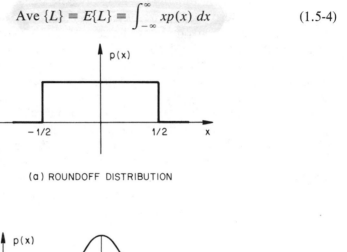

(b) GAUSSIAN DISTRIBUTION

$$p(x) = \frac{1}{\sigma \sqrt{2\pi}} e^{-\frac{(x-\mu)^2}{2\sigma^2}}$$

FIGURE 1.5-1

You are weighting each x by its probability of occurring, $p(x)$, and combining them all together. For the roundoff example,

$$\int_{-\infty}^{\infty} xp(x)\, dx = \int_{-1/2}^{1/2} x\, dx = 0$$

Note that the averaging is over the ensemble $p(x)$.
For the gaussian example (1.5-2),

$$E\{L\} = \frac{1}{\sigma\sqrt{2\pi}} \int_{-\infty}^{\infty} xe^{-1/2[(x-\mu)/\sigma]^2}\, dx$$

Replacing $(x - \mu)/\sigma$ by the normalized variable t we have

$$E\{L\} = \sigma \int_{-\infty}^{\infty} te^{-(1/2)t^2} \frac{dt}{\sqrt{2\pi}} + \mu \int_{-\infty}^{\infty} e^{-(1/2)t^2} \frac{dt}{\sqrt{2\pi}}$$

$$= 0 + \mu = \mu \tag{1.5-5}$$

The last equation is true because the first integrand is odd and therefore the integral equals zero. The second integral is $P\{-\infty < L < \infty\} = 1$.

We can think of the expectation as an operator $E\{\ \}$ operating on a function. A moment's thought and it is obvious that the expected value of a constant is the same constant,

$$E\{a\} = a$$

Again, if x is the variable of the model and if a and b are constants, then

$$E\{ax + b\} = aE\{x\} + b$$

Other "typical values" besides the average are widely used. One is the *mode*, the most frequent value or the one with maximum probability density. Another is the *median*, the value exceeded by half the distribution. We will not use them in this book.

Another commonly computed characteristic of a random quantity or, alternatively, of its distribution is the *variance*. It is denoted Var $\{L\}$ if L is the random quantity, and is defined by

$$\text{Var } \{L\} = \int_{-\infty}^{\infty} (x - \mu)^2 p(x)\, dx \tag{1.5-6}$$

where L has density $p(x)$ and $\mu = E\{L\}$. It is also denoted by the symbol σ^2. Note that the variance is always measured about the mean. In mechanics this same expression is known as the *moment of inertia*.

For the roundoff case (1.5-3), we have

$$\sigma^2 = \int_{-\infty}^{\infty} (x - 0)^2 p(x) \, dx = \int_{-1/2}^{1/2} x^2 \, dx = \frac{1}{12} \qquad (1.5\text{-}7)$$

and for the gaussian (1.5-2),

$$\sigma^2 = \text{Var} \{L\} = \frac{1}{\sigma\sqrt{2\pi}} \int_{-\infty}^{\infty} (x - \mu)^2 e^{-1/2[(x-\mu)/\sigma]^2} \, dx$$

If we set

$$x - \mu = \sigma t$$

we obtain

$$\text{Var} \{x\} = \frac{\sigma^2}{\sqrt{2\pi}} \int_{-\infty}^{\infty} t^2 e^{-t^2/2} \, dt$$

Integration by parts using

$$\begin{cases} te^{-t^2/2} \, dt = dV, & V = -e^{-t^2/2} \\ U = t, & dU = dt \end{cases}$$

gives

$$\text{Var} \{x\} = \frac{\sigma^2}{\sqrt{2\pi}} \left[-e^{-t^2/2} t \, \Big|_{-\infty}^{\infty} + \int_{-\infty}^{\infty} e^{-t^2/2} \, dt \right]$$

The integrated piece vanishes at both ends, and in the integral $\mu = 0$, $\sigma^2 = 1$. Therefore, since the integral equals $\sqrt{2\pi}$, we have

$$\text{Var} \{x\} = \frac{\sigma^2}{\sqrt{2\pi}} \sqrt{2\pi} = \sigma^2$$

This result shows why we adopted the peculiar form for writing the gaussian distribution: σ^2 is the variance and μ is the average of the gaussian (normal) distribution,

$$p(x) = \frac{1}{\sigma\sqrt{2\pi}} e^{-(x-\mu)^2/2\sigma^2} \equiv N(\mu, \sigma^2)$$

It is clear that the variance, which is the sum of the squares of the deviations of the distribution from its average value [weighted by the probability $p(x)$ of occurring], is closely related to the principle of least squares (which states that the *best fit* occurs when the sum of the squares of the errors is minimum). In both cases, it is the sum of the squares of the differences that is used. For the variance, it is the difference from the mean that is used; for a least-squares fit, it is the difference of the data from the approximate fit that is used.

Exercises

1.5-1   If the distribution for $p(x)$ is

$$p(x) = \begin{cases} 0, & x < 0 \\ ae^{-ax}, & x \geq 0 \end{cases} \quad (a > 0)$$

show that $\mu = 1/a$ and $\sigma^2 = 1/a^2$.

1.5-2   If the distribution for $p(x)$ is

$$p(x) = \begin{cases} 1 - \dfrac{x}{2}, & 0 \leq x \leq 2 \\ \\ 0, & \text{otherwise} \end{cases}$$

show that $\mu = \frac{2}{3}$, $\sigma^2 = \frac{2}{9}$.

1.5-3   Find the mean and variance of

$$p(x) = \begin{cases} \cos 2x, & -\dfrac{\pi}{4} \leq x \leq \dfrac{\pi}{4} \\ \\ 0, & \text{otherwise} \end{cases}$$

1.5-4   For a well-balanced die (singular of dice), calculate the mean and variance of the value on the top face after a random toss. Do the same for a pair of dice.

1.6   THE DISTRIBUTION OF A STATISTIC

We now turn to what is probably the hardest concept for the beginner in statistics to master, the idea of the distribution of a statistic (such as the mean or the variance of a sample).

Suppose that we have made a set of measurements and that from the sample we have computed one or more statistics. For instance, we may have selected 1000 Americans at random from the entire population of around 200 million and measured their heights. From these heights we can compute the *sample average* $\bar{x}$.

$$\bar{x} = \frac{1}{n} \sum_{i=1}^{n} x_i$$

The *variance s^2 of the sample* is defined by

$$s^2 \equiv \frac{1}{n-1} \sum_{i=1}^{n} (x_i - \bar{x})^2$$

For clarity, it is customary to use Greek letters for the statistics of the model and Latin letters for the corresponding statistics of the sample.

It is good to know these two numbers for the sample that we drew, but if we are to make much use of them, the question immediately arises: If we repeat the whole process again, using a different random sample of 1000 Americans, what could we reasonably expect to get for the average? In short, what is the distribution of the statistic called the "average"? Clearly, repetitions of the whole process of selecting the people, making the measurements, and computing the average will give us a distribution of values for the average $\bar{x}$ (and a distribution for the variance s^2).

In the roundoff example we had a unique model for the basic population from which the roundoffs were drawn, but in the gaussian example we must estimate the two unknown parameters of the population distribution μ and σ^2 from the sample statistics $\bar{x}$ and s^2. We can ask what relation these two sets of numbers have to each other. In textbooks on statistics it is proved that, for any distribution, the *average* of the sample is an unbiased estimate of the original population average. Similarly, the sample variance s^2 is an unbiased estimate of σ^2. Unbiased means that, on the average, your estimates are neither too high nor too low. (That is, the average of the statistic equals the value being estimated.)

TABLE 1.6-1 Relation of sample to population
statistics

	Sample	Population
$\text{Ave}(x) = \dfrac{1}{n} \sum_{i=1}^{n} x_i$		μ
$s^2 = \dfrac{1}{n-1} \sum_{i=1}^{n} (x_i - \bar{x})^2$		σ^2

If the sample is at all large ($n \geq 10$), then the central limit theorem shows that the statistic called the average has a distribution that is very close to a gaussian (normal) distribution

$$p(x) = \frac{\sqrt{n}}{\sigma\sqrt{2\pi}} e^{-n(x-\bar{x})^2/2\sigma^2}$$

with parameters $\bar{x}$ and σ^2/n.

Exercise

1.6-1   For the set of measurements 10, 11, 10, 12, 9, 10, 7, 10, 10, 9, compute the mean and variance of the sample and estimate the corresponding population parameters.

1.7   NOISE AMPLIFICATION IN A FILTER

Suppose that we make some measurements. Let u_n be the "true" measurement with added noise ϵ_n whose expected value is zero. The condition of zero mean is

$$E\{\epsilon_n\} = 0$$

and implies that there is no *bias* in the measurements, only local, random errors. The averaging is, of course, over the ensemble of noise ϵ_n. Furthermore, let this noise ϵ_n have a variance σ^2. What is the corresponding noise in the output of a nonrecursive filter (assuming that the arithmetic we do does not increase the noise)? To compute this, we make the additional assumption (which is often, but not always, true) that in making the measurements $u_n + \epsilon_n$ the errors ϵ_n are *uncorrelated*. This assumption in mathematical notation is

$$E\{\epsilon_n\epsilon_m\} = \begin{cases} \sigma^2, & m = n \\ 0, & m \neq n \end{cases}$$

Again the averaging is over the ensemble of the noise.
   A nonrecursive filter is defined by the formula

$$y_n = \sum_{k=-N}^{N} c_k(u_{n-k} + \epsilon_{n-k}) \tag{1.7-1}$$

The expected value is, therefore, since the E operation applies only to the ϵ_n and not to the c_k or the u_{n-k}, and $E\{\epsilon_n\} = 0$,

$$E\{y_n\} = \sum_{k=-N}^{N} c_k(u_{n-k} + E\{\epsilon_{n-k}\}) = \sum_{k=-N}^{N} c_k u_{n-k}$$

For the variance calculation, we begin with

$$E\left\{\left[\underbrace{\sum_{k=-N}^{N} c_k(u_{n-k} + \epsilon_{n-k})}_{y_n} - E(y_n)\right]^2\right\}$$

But this is (using different summation indices to keep things clear)

$$E\left\{\left[\sum_{k=-N}^{N} c_k\epsilon_{n-k}\right]^2\right\} = E\left\{\left[\sum_{k=-N}^{N} c_k\epsilon_{n-k}\right]\left[\sum_{m=-N}^{N} c_m\epsilon_{n-m}\right]\right\}$$

Since $E\{\epsilon_n\} = 0$ and, for $m \neq n$, $E\{\epsilon_n\epsilon_m\} = 0$, multiplying out and applying the operator E to the ϵ_n leaves only the terms

$$\sum c_k^2 E\{\epsilon_k^2\} = \sum c_k^2\sigma^2 = \sigma^2 \sum_{k=-N}^{N} c_k^2 \qquad (1.7\text{-}2)$$

Thus the sum of the squares of the coefficients of a filter measures the noise amplification of the filtering process. It is for this reason that the sum of the squares of the coefficients of a nonrecursive filter plays a significant role in the theory.

Exercises

1.7-1   Apply formula (1.7-2) using the roundoff noise model.

1.7-2   What is the noise amplification of the least-squares cubic (1.1-4) of Section 1.1?

1.7-3   What is the noise amplification of smoothing by 5s? *Answer:* $\sigma^2 = \frac{1}{5}$.

1.7-4   Show that the minimum noise amplification of a five-term nonrecursive filter, with the sum of the coefficients equal to 1, is the smoothing by 5s. *Hint:* Use Lagrange multipliers.

1.8 GEOMETRIC PROGRESSIONS

We need to review the topic of geometric progressions since it will occur often in the text.

A *geometric progression* is a sequence of n values

$$a, az, az^2, \ldots, az^{n-1} \tag{1.8-1}$$

where each is obtained from the previous one by multiplying by the constant z. The sum of all the terms is

$$S(n) = a(1 + z + z^2 + \cdots + z^{n-1}) \tag{1.8-2}$$

If we multiply this equation by z and then subtract it from the original we get

$$S(n) - zS(n) = a - az^n$$

$$S(n) = \frac{a(1 - z^n)}{1 - z} \tag{1.8-3}$$

The special case of symmetrically arranged terms beginning at $a = z^{-m}$ and going on for $2m + 1$ terms gives

$$z^{-m} + z^{-m+1} + \cdots + 1 + \cdots z^{m-1} + z^m$$

This sum we will label as $S(-m, m)$ and is, from (1.8-3),

$$S(-m, m) = \frac{z^{-m}(1 - z^{2m+1})}{1 - z} = \frac{z^{-m} - z^{m+1}}{1 - z}$$

Multiply the numerator and denominator by $-z^{-1/2}$

$$S(-m, m) = \frac{z^{m+1/2} - z^{-(m+1/2)}}{z^{1/2} - z^{-1/2}} \tag{1.8-4}$$

In this derivation z may be a real or complex number. Differentiating (1.8-2) and (1.8-3) with respect to z we get $(a = 1)$

$$1 + 2z + 3z^2 + \cdots + (n - 1)z^{n-2} = \frac{1 - nz^{n-1} + (n - 1)z^n}{(1 - z)^2}$$

Multiply by z to get the convenient formula

$$z + 2z^2 + 3z^3 + \cdots + (n - 1)z^{n-1}$$

$$= z \frac{[1 - nz^{n-1} + (n - 1)z^n]}{(1 - z)^2} \qquad (1.8\text{-}5)$$

If $|z| < 1$ then letting $n \to \infty$ (1.8-3) becomes

$$S(\infty) = \frac{a}{1 - z} \qquad (1.8\text{-}6)$$

and (1.8-5) becomes

$$\sum_{k=0}^{\infty} kz^k = \frac{z}{(1 - z)^2} \qquad (1.8\text{-}7)$$

We often want $1/z$ in place of z. Then (1.8-3) becomes

$$a(1 + 1/z + 1/z^2 + \cdots + 1/z^{n-1}) = \frac{az(1 - z^{-n})}{z - 1} \qquad (1.8\text{-}8)$$

and (1.8-6) is

$$a \sum_{k=0}^{\infty} z^{-k} = \frac{az}{z - 1} \qquad (1.8\text{-}9)$$

while (1.8-7) is

$$\sum_{k=0}^{\infty} kz^{-k} = \frac{z}{(z - 1)^2} \qquad (1.8\text{-}10)$$

Similar formulas are easily found by similar methods.

Exercises 1.8

1.8-1   Discuss (1.8-4) when z is replaced by $1/z$.

1.8-2   From (1.8-10) find

$$\sum_{k=0}^{\infty} k^2 z^{-k}$$

1.8-3   In (1.8-5) replace z by $1/z$.

2

The Frequency Approach

2.1 INTRODUCTION

The purpose of this chapter is to show, for linear digital filter design, why and in what sense the use of sines and cosines of the independent variable t is preferable to the classical use of polynomials in t. The approximation of a function by a polynomial is generally emphasized in mathematics, statistics, and numerical analysis. For instance, in Newton's method for finding a zero of a function $g(t)$, the function is locally replaced by the tangent line, a linear equation in t. Again, a Taylor's expansion of a function expresses the function in powers of $t - t_0$. In statistics data is constantly being fitted by polynomials. In the trapezoid rule for integration the function is locally replaced by a straight line. It is natural, therefore, to suppose that in other fields polynomials are the proper functions to use when approximating a given function. Thus we are concerned in this chapter more with the psychological problem of undoing this earlier conditioning in favor of polynomials than with the logical problem of presenting the frequency approach.

Before doing this, however, we introduce in the next section the most important consequence of sampling a function at equally spaced points. This phenomenon, called *aliasing*, is a common experience for most people; but they are so accustomed to it that they are only vaguely aware of it.

We shall then show, in three different senses, that the sines and cosines are the proper functions for situations that are relevant to much of data processing on computers. To do so, it is necessary to introduce the concepts of *eigenfunctions* and *eigenvalues* and to show that the concept of the *transfer function* corresponds to the eigenvalues of the process.

Since the idea of *frequency* is clearly central to the frequency approach,

21

we must be careful to say what it means. Consider, for example, a rectangular wave (or any other shaped wave) that exactly repeats itself 10 times a second; we say that it has a *period* (cycle) of $T = \frac{1}{10}$ second and a *rotational frequency* of 10 hertz (cycles per second). Hertz is abbreviated Hz when used as a unit of measure. By a period of a function, we mean the *shortest* interval for which the function exactly repeats itself, and the *fundamental frequency* is the corresponding frequency.

The period T and the frequency f are reciprocals of each other. The *angular frequency* ω (in radians) is related to the rotational frequency f by

$$\omega = 2\pi f \qquad (2.1\text{-}1)$$

The use of the angular measure ω is natural in calculus situations, and use of the rotational frequency f is natural in applications. This is the way that we will use them. This use of two different units occurs similarly in logs; in the theory part, as in the calculus, we use natural logs, while in practice we use the logs to the base 10.

The adjective *fundamental* is often dropped, but doing so can lead to confusion. For instance, in Section 4.3 we will decompose a rectangular wave into a sum of sines and cosines and then say that the original wave form has high frequencies in it. Thus confusion can arise concerning the frequency of the original wave form and the frequencies of the terms in the decomposition of the wave into a set of sinusoidal (periodic) functions.

2.2 ALIASING

The phenomenon of *aliasing*, which is basic to sampling data at equally spaced intervals, is not new to the reader who has watched Westerns either on television or in the movies. As the stagecoach wheels turn faster and faster, they appear to slow down and then to stop. If the increase in speed is great enough, they may seem to go backward, stop, and go forward a number of times. Any actual high rate of rotation of the wheels appears, as a result of the equally spaced *sampling of the pictures* in time, to be "aliased" into a low frequency of rotation. Figure 2.2-1 shows, symbolically, a wheel with four spokes rotating at different rates, and the human mind interprets what is seen as the smallest motion that accounts for the observations. At the first time that the wheel seems to stop, it will appear to have twice the normal number of spokes.

Another common application of this phenomenon of *aliasing due to sampling* occurs when a stroboscope is flashed at a rate close to that of a piece of rotating equipment. If the stroboscope flashes at a rate slightly *less* than the rate of rotation (or some multiple of it), then the flashes make the

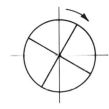

SLOW ROTATION	MIDDLE ROTATION	FASTER ROTATION
APPEARS AS A DIRECT ROTATION	APPEARS AS IF STAND-ING STILL WITH DOUBLE THE NUMBER OF SPOKES	APPEARS AS A BACK-WARD ROTATION

FIGURE 2.2-1

machine appear to the eye as if it were rotating slowly forward; the closer the rates, the slower the apparent rotation. Again we see that one frequency is aliased into another due to the process of taking equally spaced samples.

In the case of a sinusoid, we are not sampling a rotating wheel but are, in effect, sampling one component, either vertical or horizontal (or in any other direction for that matter). As a result, we can see the phenomenon of aliasing due to sampling at equal intervals in time (the independent variable) as a simple consequence of trigonometric identities.

Consider the sinusoid

$$u(t) = \cos[2\pi(m + a)t + \phi]$$

where m is an integer, positive or negative, a is the *positive* fractional part of the original rate of rotation, and ϕ is an arbitrary phase angle. Since we are sampling the sinusoid at integer values of t, the reduction of any angle by $2m\pi$ leaves the cosine with the same values at the sample points. Therefore, the sinusoid is equivalent (at the sample points $t_n = n$) to

$$\cos[2\pi a t + \phi]$$

If $a > \frac{1}{2}$, we can remove another 2π, and [since $\cos x = \cos(-x)$]

$$\cos[2\pi(-1 + a)t + \phi] = \cos[2\pi(1 - a)t - \phi]$$

at the sample points. Thus we have shown, using only simple trigonometry, that *at the sample points* any sinusoid of arbitrary frequency is equivalent to a sinusoid with a frequency that lies between 0 and $\frac{1}{2}$, equivalent in the sense that the two sinusoids have the same numerical values at the sample points (see Figure 2.2-2). In a very real sense, the two frequencies are indistinguishable: a high frequency is aliased into (appears as) a low frequency

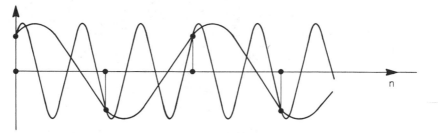

FIGURE 2.2-2 ALIASING

due solely to the sampling process. Only sinusoids with frequencies low enough so that at least two samples occur in each period are not aliased.

To illustrate, the function

$$u_n = \cos \frac{7\pi}{2} n \qquad (n = 0, \pm 1, \pm 2, \ldots)$$

has the same values as

$$u_n = \cos\left(\frac{7\pi}{2} - 4\pi\right)n = \cos\left(\frac{-\pi}{2} n\right) = \cos \frac{\pi}{2} n$$

at all the sample points, and hence the original function is aliased into the lower-frequency function.

Exercises

2.2-1   A machine rotates at 100 Hz. If a strobe light flashes at a rate of 99 per second, what is the apparent motion of the machine? (*Hint:* Take $\frac{1}{99}$ second as the unit of time for the sampling.) If the strobe flashes 101 times per second? 98 times?

2.2-2   Find the lowest aliased frequency of $\cos[8\pi n/3 + \pi/3]$. Of $\cos[13\pi n/3 + \pi/3]$.

2.2-3   Repeat the argument for sines in place of cosines. Note carefully the small differences between the cosines and sines.

2.2-4   What is the aliased frequency of $\cos 4\pi n$?

2.2-5   What is the highest frequency of a constant?

2.2-6   A compact disc recording uses 44,100 samples per second. What is the highest unaliased frequency? *Answer:* 22,050

2.3 THE IDEA OF AN EIGENFUNCTION

The word *eigenfunction* is a half-translation from the German of what in the older English texts was called characteristic function, proper function, or natural function.

For purposes of illustration *only* consider a special case of eigenfunctions, the multiplication of a square matrix $\mathbf{A} = (a_{i,j})$, of dimension N by N, by a vector $\mathbf{x}$, of dimension N by 1. The product is another vector $\mathbf{y}$, of dimension N by 1,

$$\mathbf{Ax} = \mathbf{y}$$

If $\mathbf{A}$ is the identity matrix, then, of course, the vector $\mathbf{x}$ equals the vector $\mathbf{y}$ in the sense that all the components have the same value. Also, if $\mathbf{x} = 0$, then $\mathbf{y} = 0$, but in the future we shall exclude the function (vector) that is identically zero.

Usually the output vector $\mathbf{y}$ will point in a different direction (in the N-dimensional space) from the input vector $\mathbf{x}$. For the typical matrix $\mathbf{A}$ of dimension N, there will be N different vectors $\mathbf{x}$ such that the corresponding $\mathbf{y}$ will have the *same direction* as did the $\mathbf{x}$, although not necessarily the same length. That is, we will have

$$\mathbf{Ax} = \lambda\mathbf{x}$$

for some constant λ. To see the truth of this remark, we can write the preceding equation in the form

$$(\mathbf{A} - \lambda\mathbf{I})\mathbf{x} = 0$$

where $\mathbf{I}$ is the identity matrix. For this equation to have a solution that is not identically zero, it is both necessary and sufficient that the determinant of the system of equations

$$|\mathbf{A} - \lambda\mathbf{I}| = 0$$

This determinant,

$$\begin{vmatrix} a_{11} - \lambda & a_{12} & a_{13} & \cdots \\ a_{21} & a_{22} - \lambda & a_{23} & \cdots \\ a_{31} & a_{32} & a_{33} - \lambda & \cdots \\ \vdots & \vdots & \vdots & \end{vmatrix} = 0$$

when expanded, is clearly a polynomial in λ of degree N, and, in general, it will have N distinct zeros $\lambda_1, \lambda_2, \ldots, \lambda_N$, real or complex (to have a multiple zero is a restriction on the matrix $\mathbf{A}$). Thus there are, in general, N distinct λ_k with corresponding vector solutions $\mathbf{x}_k$. (Note that $\mathbf{x}_k$ is a vector and not a component of a vector.) The values λ_k are called the *eigenvalues*, and the $\mathbf{x}_k$ are called the corresponding *eigenvectors*. For a given eigenvalue, the determinant is zero, and the corresponding eigenvector is, of course, determined only to within a multiplicative constant.

Why are these eigenvectors important? There are (almost always) N distinct eigenvectors, and they can be shown to be *linearly independent*. Therefore, they can serve as a basis for representing an arbitrary vector $\mathbf{x}$ of N dimensions. Thus we can represent an arbitrary vector $\mathbf{x}$ as a linear combination of the N eigenvectors $\mathbf{x}_k$:

$$\mathbf{x} = \sum_{k=1}^{N} a_k \mathbf{x}_k$$

If we now multiply this equation on both sides by the matrix $\mathbf{A}$ (technically, we apply the operation $\mathbf{A}$ to the equation), we find that

$$\mathbf{A}\mathbf{x} = \sum_{k=1}^{N} a_k \mathbf{A}\mathbf{x}_k = \sum_{k=1}^{N} a_k \lambda_k \mathbf{x}_k$$

and we see that each eigenvector is multiplied by its corresponding eigenvalue. In the eigenvector representation, the effect of the multiplication by the matrix $\mathbf{A}$ (applying the operation $\mathbf{A}$) is easy to follow. The eigenvectors are independent of each other.

To state this important property in other words, we can say that the eigenfunctions do not "feel" the presence of each other; each minds its own business regardless of how much there may or may not be of the other eigenfunctions.

To illustrate the above consider the matrix

$$\mathbf{A} = \begin{pmatrix} 1 & 2 \\ 3 & 2 \end{pmatrix}$$

This leads to the corresponding determinant

$$|\mathbf{A} - \lambda\mathbf{I}| = \begin{vmatrix} 1 - \lambda & 2 \\ 3 & 2 - \lambda \end{vmatrix} = 0$$

Upon expansion of the determinant, we get the equation for the eigenvalues

$$\lambda^2 - 3\lambda + 2 - 6 = 0$$

which has zeros $\lambda = 4, -1$. If we use $\lambda_1 = 4$, we obtain for the matrix equation (where we are using the notation $x_{i,j}$ for the jth component of the ith vector $\mathbf{x}_i$)

$$\begin{pmatrix} -3 & 2 \\ 3 & -2 \end{pmatrix} \begin{pmatrix} x_{1,1} \\ x_{1,2} \end{pmatrix} = 0$$

which leads to the single equation

$$-3x_{1,1} + 2x_{1,2} = 0$$

$$x_{1,2} = \frac{3\,x_{1,1}}{2}$$

and the corresponding eigenvector

$$\begin{pmatrix} x_{1,1} \\ \frac{3x_{1,1}}{2} \end{pmatrix} = \frac{x_{1,1}}{2} \begin{pmatrix} 2 \\ 3 \end{pmatrix}$$

The value of $x_{1,1}$ is arbitrary, since the rank of the matrix for the eigenvalue is 1. If we use the other eigenvalue $\lambda_2 = -1$, we obtain, correspondingly,

$$\begin{pmatrix} 2 & 2 \\ 3 & 3 \end{pmatrix} \begin{pmatrix} x_{2,1} \\ x_{2,2} \end{pmatrix} = 0$$

with the corresponding eigenvector

$$\begin{pmatrix} x_{2,1} \\ -x_{2,1} \end{pmatrix} = x_{2,1} \begin{pmatrix} 1 \\ -1 \end{pmatrix}$$

These two eigenvectors,

$$\begin{pmatrix} 2 \\ 3 \end{pmatrix} \quad \text{and} \quad \begin{pmatrix} 1 \\ -1 \end{pmatrix}$$

are linearly independent and can represent any arbitrary two-dimensional vector.

In particular the general vector

$$\begin{pmatrix} a \\ b \end{pmatrix} = \frac{a + b}{5} \begin{pmatrix} 2 \\ 3 \end{pmatrix} + \frac{3a - 2b}{5} \begin{pmatrix} 1 \\ -1 \end{pmatrix}$$

hence

$$\mathbf{A} \begin{pmatrix} a \\ b \end{pmatrix} = 4 \frac{a + b}{5} \begin{pmatrix} 2 \\ 3 \end{pmatrix} - \frac{3a - 2b}{5} \begin{pmatrix} 1 \\ -1 \end{pmatrix}$$

Exercises

2.3-1   Find the eigenvalues and eigenvectors of the matrix

$$\mathbf{A} = \begin{pmatrix} 1 & -1 \\ 2 & 0 \end{pmatrix}$$

2.3-2   Given the matrix

$$\mathbf{A} = \begin{pmatrix} 1 & 1 & 1 \\ 1 & 0 & -2 \\ 1 & 0 & 0 \end{pmatrix}$$

find all the eigenvalues and the eigenvector corresponding to the eigenvalue 1. *Answer:* $(1, -1, 1)$.

2.4   INVARIANCE UNDER TRANSLATION

In many data-processing problems there is no natural origin, and therefore an arbitrary point is selected as the origin (typically, for a time signal, it is the time when we set $t = 0$, which is arbitrary). From the addition formulas of trigonometry

$$\sin(x + y) = \sin x \cos y + \cos x \sin y$$
$$\cos(x + y) = \cos x \cos y - \sin x \sin y \qquad (2.4\text{-}1)$$

it is an easy exercise in trigonometry to see that, when $x = x' + h$,

$$A \sin x + B \cos x$$

becomes

$$A' \sin x' + B' \cos x'$$

where we have

$$A' = A \cos h - B \sin h$$
$$B' = A \sin h + B \cos h \qquad (2.4\text{-}2)$$

Squaring each of these expressions and adding them, we obtain

$$A'^2 + B'^2 = A^2 + B^2 \qquad (2.4\text{-}3)$$

Thus we see that under the operation of translation the *pair* of functions cos x and sin x constitutes the eigenfunctions, for when placed into the operation of translation by the amount h, they again emerge.

The famous Euler identities are

$$\cos x + i \sin x = e^{ix}$$
$$\cos x - i \sin x = e^{-ix} \tag{2.4-4}$$

(where $i = \sqrt{-1}$). Note that, for x real, as is assumed,

$$| e^{ix} | = 1 \tag{2.4-5}$$

To verify the equations (2.4-4) we expand e^{ix} in a power series

$$e^{ix} = 1 + (ix) + \frac{(ix)^2}{2!} + \frac{(ix)^3}{3!} + \frac{(ix)^4}{4!} + \frac{(ix)^5}{5!} + \dots$$

Rearrange and separate the real and imaginary parts, noting that $i^2 = -1$, $i^4 = 1$, etc.

$$e^{ix} = \left(1 - \frac{x^2}{2!} + \frac{x^4}{4!} - \dots\right) + i\left(x - \frac{x^3}{3!} + \frac{x^5}{5!} - \dots\right)$$

and we recognize the corresponding sine and cosine expansions. Thus we have the top equation of (2.4-4). Replace i by $-i$ and you have the lower equation.

The equations (2.4-4) lead to the corresponding formulas

$$\cos x = \frac{1}{2}(e^{ix} + e^{-ix})$$

$$\sin x = \frac{1}{2i}(e^{ix} - e^{-ix}) \tag{2.4-6}$$

In this notation the two addition formulas (2.4-1) from trigonometry are contained in the single, much simpler formula

$$e^{ix}e^{iy} = e^{i(x+y)}$$

This fact can be seen by using the Euler identities (2.4-4) on both sides and then equating the real and imaginary terms on each side. Thus the complex

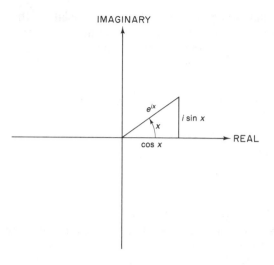

exponentials are the eigenfunctions of translation. *It is far more convenient, therefore, to use the complex exponentials than it is to use the real sines and cosines.* On the other hand, the real functions are familiar, and the complex exponentials have a mysterious aura about them. It is simply necessary to get used to the fact that the complex exponentials are the real functions in a slight disguise. You can think of the cosine and sine as the components of a vector e^{ix}. See Figure 2.4-1.

We have chosen the mathematical convention of $i = \sqrt{-1}$ rather than the engineering convention of j. The choice is arbitrary, but since the book is designed to be read by nonengineers, i is a reasonable choice, and engineers need to be familiar with both notations.

It is natural to ask if the sine and cosine are unique in having this property of invariance under translation. The invariant property that we want is that both functions under a translation of a fixed amount, say h, can be written as a linear combination of sine and cosine or, more generally, that any linear combination of a sine and cosine of a given frequency can be written as a linear combination when an arbitrary translation of size h of the coordinate axis is made. We are further assuming that the functions are odd and even, respectively, and that the trigonometric functions, sine and cosine, are reasonably smooth. Under these assumptions it can be shown that the sines and cosines are unique (except that the corresponding hyperbolic functions are also possible). Notice that in the complex exponential notation the eigenfunction property is much more simply expressed than in the trigonometric notation,

$$u(t + h) = e^{i\omega(t+h)} = e^{i\omega h}e^{i\omega t} = \lambda(\omega)u(t)$$

where $\lambda(\omega)$ is the eigenvalue

$$\lambda(\omega) = e^{i\omega h}$$

and is independent of the variable t.

Thus the $e^{i\omega t}$ are the eigenfunctions of translation.

Corresponding to equation (2.4-1), we have in the complex notation

$$u(t) = ce^{i\omega t}$$

(where c can be a complex number) the invariant (2.4-3)

$$(ce^{i\omega t})(\bar{c}e^{-i\omega t}) = c\bar{c} = |c|^2 \qquad (2.4-7)$$

2.5 LINEAR SYSTEMS

The second eigenfunction property that we wish to show is that the complex exponential functions $e^{i\omega t}$ and $e^{-i\omega t}$ are eigenfunctions for linear, time-invariant systems. In abstract notation this means

$$L\{e^{i\omega t}\} = \lambda(\omega)e^{i\omega t}$$

where $L\{\ \}$ is an arbitrary linear time invariant operator and $\lambda(\omega)$ does not depend on t. A linear operator has the property that

$$L\{ag_1(t) + bg_2(t)\} = aL\{g_1(t)\} + bL\{g_2(t)\}$$

For example, integration, differentiation, and interpolation are all linear operators. Clearly, for nonrecursive filters of the form

$$y_n = \sum_{k=-N}^{N} c_k u_{n-k}$$

the substitution

$$u(t) = e^{i\omega t}$$

produces, when we factor out the exponential term depending on n, the output

$$y(n) = e^{i\omega n} \sum_{k=-N}^{N} c_k e^{-i\omega k} = \lambda(\omega)e^{i\omega n} \qquad t = n - k.$$

where

$$\sum_{k=-N}^{N} c_k e^{-i\omega k} = \lambda(\omega) \tag{2.5-1}$$

Thus the function $e^{i\omega t}$, which we put into the right side of the equation, can be factored out of the expression and appears multiplied by its eigenvalue $\lambda(\omega)$. The eigenvalue $\lambda(\omega)$ is, of course, a constant as far as t or, equivalently, n is concerned, and is usually called the *transfer function*. Thus the transfer function is just the eigenvalue corresponding to the eigenfunction $e^{i\omega t}$.

For a recursive filter we need to assume that *both* the input and the output are of the form

$$u_n = A_I e^{i\omega n} \quad \text{and} \quad y_n = A_O e^{i\omega n}$$

where the input coefficient A_I and the output coefficient A_O may be complex numbers. We get from equation (1.1-5)

$$A_O e^{i\omega n} = A_I \sum_0^N c_k e^{i\omega(n-k)} + A_O \sum_1^M d_k e^{i\omega(n-k)}$$

or the ratio of output to input coefficients

$$\frac{A_O}{A_I} = \frac{\sum_0^N c_k e^{-i\omega k}}{1 - \sum_1^M d_k e^{-i\omega k}} \tag{2.5-2}$$

This is a transfer function because, when you multiply the input by this ratio you get the output; you "transfer" from the input to the output.

It is worth noting that the exponential function is also the eigenfunction that is appropriate for the calculus operations of differentiation,

$$\frac{d}{dt} e^{i\omega t} = i\omega e^{i\omega t} \tag{2.5-3}$$

and integration,

$$\int e^{i\omega t} \, dt = \frac{e^{i\omega t}}{i\omega} \tag{2.5-4}$$

The exponential is also the eigenfunction for differencing, since

$$\Delta e^{i\omega t} = e^{i\omega(t+1)} - e^{i\omega t} = e^{i\omega t}[e^{i\omega} - 1]$$

Thus we see, contrary to the impression gained from the usual calculus course, that the powers of x are not the eigenfunctions of calculus. Instead, the exponentials, real or complex, are the natural, the characteristic, the eigenfunctions of the calculus.

The expression $e^{i\omega}$ occurs frequently and it is often easier to make a notational change and write

$$e^{i\omega} = z$$

Hence (2.5-1) has the form

$$\sum_{k=-N}^{N} c_k z^{-k} = \lambda(\omega)$$

and (2.5-2) has the form

$$\frac{A_O}{A_I} = \frac{\sum\limits_{0}^{N} c_k z^{-k}}{1 - \sum\limits_{1}^{M} d_k z^{-k}}$$

Exercises

2.5-1   Find the eigenvalue corresponding to the kth derivative.

2.5-2   Find the eigenvalue corresponding to the kth difference operator Δ^k.

2.5-3   Discuss the lack of an additive constant in (2.5-4).

2.6   THE EIGENFUNCTIONS OF EQUALLY SPACED SAMPLING

The purpose of this section is to show that the eigenfunctions of the process of equally spaced sampling of a function are the common sines and cosines of trigonometry (or equally the complex exponentials e^{ix} and e^{-ix}). The sense in which we mean that they are eigenfunctions is that when we (1) take a sinusoid of some frequency (think of it as a high frequency), then

(2) do the process of sampling the given frequency function at equally spaced points, and (3) finally ask "What equivalent sinusoid of low frequency do we have?" we find that it is equivalent to a *single* sinusoid function. Stated simply, aliasing takes any particular frequency and, in the sense of having the same values at the sample points, transforms it into a single low-frequency function.

Let us contrast this result with what happens when the classical polynomial method of approximation is used. In polynomial approximation when we use the sample points x_i $(i = 1, \ldots, N)$, we are led directly to consider the *sample polynomial* defined by

$$\pi(x) = [x - x_1][x - x_2] \cdots [x - x_N]$$

This function plays a central role in the theory, because it is the function that vanishes at all the sample points x_i and thus is the function that we cannot "see." Now, given any power of x, say x^m, we divide this power by $\pi(x)$ in order to get a quotient $Q(x)$ and a remainder $R(x)$:

$$x^m = \pi(x)Q(x) + R(x)$$

where $R(x)$ is of degree less than N. A simple generalization of the standard remainder theorem shows that *at the sample points x_i* the two functions x^m and $R(x)$ have exactly the same values. Thus the original single power of x is aliased into a polynomial $R(x)$, which is, of course, a linear combination of $1, x, x^2, \ldots, x^{(N-1)}$, and not a single power. In this sense, the powers of x are not eigenfunctions for sampling at any spacing. Aliasing for polynomials is a messy business.

Let us restate this result. If we regard the process as (1) starting with a basis function (a power of x, say x^m), (2) sampling at N points, and finally, (3) constructing from the samples a new function of minimal degree in x, then we see that, in general, a single power of x does not go into a power of x. On the other hand, for sinusoids, the process of equally spaced sampling followed by the reconstruction of a function of minimal frequency does result in a single sinusoid. Consequently, in this sense, the sinusoids are the eigenfunctions of equally spaced sampling, and the process reveals once more the central role that aliasing plays in the equally spaced sampling process.

2.7 SUMMARY

In this chapter we have discussed the phenomenon of aliasing, which is due solely to the equally spaced sampling of the original signal. We have also given three reasons why the trigonometric functions, sine and cosine, are to be used in many filter problems as the basis of representing signals. The reasons are that they are the eigenfunctions for (1) invariance under translation by an arbitrary amount, (2) linear systems, and (3) equally spaced sample systems.

In the complex form $e^{i\omega t}$ and $e^{-i\omega t}$, the trigonometric functions are more easily handled in many problems. Unfortunately, most students believe that we are modeling a real world, and they believe that in the final analysis they have to deal with real signals. The complete equivalence of the two forms, real and complex, does not convince them that the two approaches are exactly equivalent.

At any frequency we have both a sine and a cosine as the basis for representation of any function; in the complex notation we have the positive and negative frequencies to use, and thus the same amount of linear independence. Ultimately, the convenience of the complex notation must be mastered, because it also leads more readily to the deeper insights of what is going on with all signal processing.

In recognition of the reality of the prejudice, we will for a time continue to give both the real and complex forms; but finally we will have to settle on the complex notation as our main tool. When you have to deal with a real function, then the coefficients in the complex form are conjugates of each other, which makes the two terms conjugates of each other, and their sum is thus real.

We have also introduced the z- transform

$$z = e^{i\omega}$$

which at present is a mere notational convenience and adds nothing to the theory.

3

Some Classical Applications

3.1 INTRODUCTION

This chapter uses the frequency approach to examine a number of classical applications of digital filters (although they are usually not presented as digital filters). Looking first at familiar situations makes the frequency approach both easier to understand and psychologically more acceptable. The original derivation in each case was based on the classical polynomial approach to numerical methods, which uses polynomials as the basis for approximation. By comparing the frequency approach to the polynomial approach, the frequency approach will be shown to shed new light on well-known situations.

Although the chapter is long, it simply represents a sequence of applications using exactly the same set of ideas. In a sense the reader is being overwhelmed with particular cases. It is hoped that a few of the examples will be sufficiently familiar so that the new approach will produce significant new insights. Thus the size of the chapter is due more to the variety of assumed backgrounds of the readers than to the importance of the individual examples. We are also familiarizing the student with the method of least squares and the exponential notation. Many of the formulas are useful in practical computing.

Having mastered this new method of analyzing and evaluating linear formulas by using complex exponentials, it is natural to turn to the corresponding design problem. To do this we will need the formal mathematics of Fourier series, which is discussed in Chapters 4 and 5.

3.2 LEAST-SQUARES FITTING
OF POLYNOMIALS

We first derive a formula for M data points (t_m, u_m), $m = 1, 2, \ldots$, M, when we wish to fit (approximate) this data, in some sense, by a polynomial $u = u(t)$ of degree N, where $M > N + 1$ (meaning that there are more data points than there are parameters in the polynomial). See Figure 3.2-1. In general, we cannot hope to find coefficients so that the polynomial exactly fits all the data (even neglecting computer roundoff). Since we assumed that the measurements are made at exactly the correct points, all the errors are in the vertical direction. The differences between the polynomial and the original data are called the *residuals*. As a rule, the residuals will not all be zero. *The principle of least squares* states that of all polynomials of degree N we should select the one for which the sum of the squares of the *residuals* is the least. This principle is, of course, an assumption and should not be taken as infallible truth.

As an example, consider a set of five equally spaced data points. For case of finding the formula, we first fix the coordinate system and pick the points at $t_m = m$ for $m = -2, -1, 0, 1, 2$, with the corresponding values of u_m unspecified. If we fit this data by a straight line $u = A + Bt$, in a least-squares sense, then we minimize

$$F(A, B) = \sum_{m = -2}^{2} [u_m - (A + Bm)]^2$$

As indicated on the left of the equation the variables of the problem are the coefficients A and B of the straight line.

u(t)

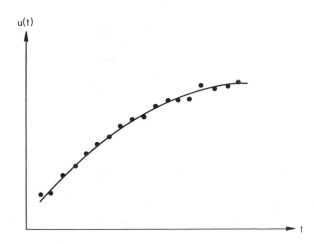

FIGURE 3.2-1 FITTING DATA LEAST SQUARES

To find a minimum, we naturally differentiate with respect to A and B and then set the resulting expressions equal to zero. We are fitting a line and the particular line we pick depends on the coefficients A and B. We have the table

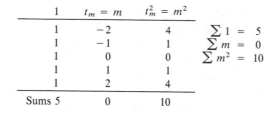

1	$t_m = m$	$t_m^2 = m^2$	
1	-2	4	$\sum 1 = 5$
1	-1	1	$\sum m = 0$
1	0	0	$\sum m^2 = 10$
1	1	1	
1	2	4	
Sums 5	0	10	

This process gives the following two equations (since $\sum t_m = \sum m = 0$ and $\sum t_m^2 = \sum m^2 = 10$):

$$
\begin{cases}
5A + 0B = \displaystyle\sum_{m=-2}^{2} u_m \\[2em]
0A + 10B = \displaystyle\sum_{m=-2}^{2} m u_m
\end{cases}
$$

These equations are called *the normal equations*, presumably because they normally arise in least-squares polynominal fitting. Before solving them for both A and B, what do we want? In the common case of smoothing, we use the midpoint of the line as the smoothed value $\bar{u}$ in place of the original data point u_0. In this case, we need find only A, which is, from the first normal equation, one-fifth of the sum of the five data values. Here the smoothed value is the mean of the five values $u_{-2}, u_{-1}, u_0, u_1, u_2$.

We fitted the straight line to the five points, but typically, when smoothing a long run of data, we fit a straight line to each (overlapping) set of five adjacent points and take the corresponding smoothed values $\bar{u}$ as the values at the midpoint of the corresponding five points. But a polynomial shifted is still a polynomial of the same degree. Therefore, when we go to the moving coordinate system to do the smoothing at successive points, we find that we are still smoothing by 5s and that we have the formula

$$
\bar{u}_n = \tfrac{1}{5} \sum_{m=-2}^{2} u_{n-m}
$$

where $\bar{u}_n$ is the smoothed value at $t = t_n$. The five coefficients c_m of the filter are each $\tfrac{1}{5}$. The noise amplification factor (equation 1.7-2) is also $\tfrac{1}{5}$. Note that we are not able, with this formula, to smooth the two values at the start of the run of data we are given and the two values at the end of the data.

The window through which we are looking at the data can be represented by its coefficients in the form

$$\tfrac{1}{5}\lfloor 1, 1, 1, 1, 1 \rfloor$$

It is a convolving window operating on the data u_n.

How does this formula we have just derived look from the frequency point of view? Suppose that the input function is one of the eigenfunctions in the complex form $e^{i\omega t}$ of frequency ω. Since the formula is linear in the data, we know (from Section 2.5) that the same function will emerge *except* that it is multiplied by its eigenvalue, which we will label from now on as

$$H(\omega)$$

The eigenvalue depends on ω but not on t (in this case, n). Putting the complex exponential $e^{i\omega t}$ into the smoothing-by-5s formula, we obtain (for $t = -2, -1, 0, 1, 2$)

$$H(\omega) = \tfrac{1}{5}[e^{-2i\omega} + e^{-i\omega} + 1 + e^{i\omega} + e^{2i\omega}]$$

Note that the negative and positive frequencies have the same coefficient, and therefore they can be replaced by the corresponding cosine functions

$$H(\omega) = \tfrac{1}{5}[1 + 2 \cos \omega + 2 \cos 2\omega]$$

This function $H(\omega)$, which is the eigenvalue of the formula, is called *the transfer function*, since it is what enables us to take the input frequency and *transfer* to the output by simply multiplying by $H(\omega)$. This curve, along with several to be discussed in the following paragraphs, is shown in Figure 3.2-2 and is labeled 5 point. Here we have used frequency f as the independent variable, where $f = \omega/2\pi$.

An alternate representation of the transfer function is based on the fact that in the complex exponential form it is a geometric progression with $r = e^{i\omega}$; hence

$$H(\omega) = \frac{e^{-2i\omega} + e^{-i\omega} + 1 + e^{i\omega} + e^{2i\omega}}{5}$$

Set $e^{i\omega} = z$ and use (1.8-4) and (2.4-6) to get

$$H(\omega) = \frac{e^{5i\omega/2} - e^{-5i\omega/2}}{5[e^{i\omega/2} - e^{-i\omega/2}]} = \frac{\sin(5\omega/2)}{5 \sin(\omega/2)}$$

The transfer function, $H(\omega)$, is clearly a periodic function of ω. However, because of the original sampling at unit spacing, it has meaning only in an interval of length 2π, which we conventionally take as running from $-\pi$ to π. To extend the transfer function beyond this range is to confuse frequencies that are already aliased by the sampling into the fundamental interval with those that before the sampling lay outside. Thus we shall always draw at most one period $-\frac{1}{2}$ to $\frac{1}{2}$ in f of the transfer function, and usually only from 0 to $\frac{1}{2}$. This cutoff frequency, where the aliasing due to the sampling process occurs, is called the *folding* or *Nyquist frequency*.

For convenience, we always plot our figures in terms of the rotational frequency f rather than the angular frequency ω. The quantities f and ω are related by the formulas

$$2\pi f = \omega \qquad \text{or} \qquad f = \frac{\omega}{2\pi}$$

Thus $H(\omega) = H(2\pi f)$. Unfortunately, this latter function is often written $H(f)$, which is a source of confusion. We will therefore write

$$H(2\pi f) = \tilde{H}(f)$$

It is the same confusion that occurs, say, with the sine function when radians and degrees are both used in the same discussion.

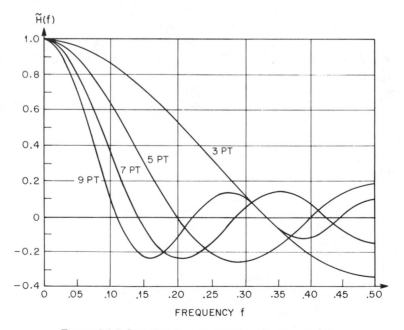

FIGURE 3.2-2 SMOOTHING BY LEAST SQUARES STRAIGHT LINES

Figure 3.2-2 shows graphically what happens at any individual frequency f: the amplitude of the input function is simply multiplied by the corresponding eigenvalue $\tilde{H}(f)$ plotted in the figure. The lowest frequency $f = 0$ (in electrical applications this corresponds to direct current and is often abbreviated dc) is transmitted through the smoothing filter with the correct amplitude in this case. All other frequencies suffer some attenuation (decrease). The smoothing by $5s$ has two frequencies, $f = \frac{2}{10}$ and $\frac{4}{10}$, such that $\tilde{H}(f)$, hence the amplitude of the output, is zero regardless of the amplitude of the corresponding input frequency.

To generalize this smoothing formula, suppose we fit a straight line to $2N + 1$ points. We proceed as we did for the five-point formula, and we are led promptly to smoothing by averaging the $2N + 1$ points. The corresponding transfer function $H(\omega)$ gives the eigenvalue for frequency ω [use (1.8-4) and (2.4-6)],

$$H(\omega) = \frac{1}{2N + 1} [e^{-iN\omega} + e^{-i(N-1)\omega} + \cdots + e^{iN\omega}]$$

$$= \frac{\sin(N + \frac{1}{2})\omega}{(2N + 1) \sin(\omega/2)} \qquad (3.2\text{-}1)$$

From this formula we see that the more terms used in the smoothing formula, the more rapid are the wiggles of the transfer function $H(\omega)$, and the more the envelope of the wiggles is squeezed toward the frequency axis. For the graphs of the first few cases, $2N + 1 = 3, 5, 7, 9$, see Figure 3.2-2. Notice that the $\sin(\omega/2)$ in the denominator goes from 0 to 1 as ω goes from 0 to π (f goes from 0 to $\frac{1}{2}$). Only the positive half of the transfer function is shown.

These smoothing formulas are clearly the same as the running *average* in statistics, and thus the transfer function $H(\omega)$ shows what happens when data of frequency ω is averaged. Notice that these transfer functions are periodic and symmetric about both $\omega = 0$ and $\omega = \pi$ ($f = \frac{1}{2}$). The periodicity about $f = \frac{1}{2}$ reflects the aliasing that we earlier found must occur when sampling at equally spaced points. The periodicity about $f = 0$ arises from the symmetry of the coefficients of the formula.

Since we are using eigenfunctions, the linear operation of smoothing does not cause the various terms to interact. Thus if the input function $u(t)$ is a sum of M complex sinusoids of frequencies ω_n and amplitudes C_m

$$u(t) = \sum_{m=1}^{M} C_m e^{i\omega_m t}$$

then the output is

$$y(t) = \sum_{m=1}^{M} C_m H(\omega_m) e^{i\omega_m t}$$

Exercises

3.2-1   Construct (real-valued) data corresponding to $f = 4/10$ and smooth it by $5s$.

3.2-2   Derive the formula for smoothing by $(2N + 1)$-point moving average.

3.2-3   Give explicit formulas for the zero crossings of the transfer functions in Figure 3.2-2 and for the values at $f = \frac{1}{2}$.

3.2-4   Examine the smoothing formula

$$y_n = \tfrac{1}{8}[u_{n-2} + 2u_{n-1} + 2u_n + 2u_{n+1} + u_{n+2}]$$

3.2-5   Examine the smoothing formula

$$y_n = \tfrac{1}{9}[u_{n-2} + 2u_{n-1} + 3u_n + 2u_{n+1} + u_{n+2}]$$

3.3   LEAST-SQUARES QUADRATICS AND QUARTICS

Instead of smoothing by fitting a straight line, suppose that we fit a quadratic (or a cubic, which is equivalent). For the quadratic

$$u(t) = A + Bt + Ct^2$$

we minimize the sum of squares of the residuals,

$$F(A, B, C) = \sum[u_m - (A + Bm + Cm^2)]^2$$

By differentiating with respect to the variables of the fit, that is, A, B, and C, we obtain the normal equations

$$\begin{cases} A\{1\} + B\{m\} + C\{m^2\} = \{u\} \\ A\{m\} + B\{m^2\} + C\{m^3\} = \{mu\} \\ A\{m^2\} + B\{m^3\} + C\{m^4\} = \{m^2u\} \end{cases}$$

where $\{\ \}$ means "sum of all the values of the quantities enclosed in the braces." By selecting the particular set of equally spaced data points $-m$, $-(m - 1), \ldots, m$, the sums over the odd powers are all zero. Solving the first and third equations for A, which is the only value we need, we have

$$A = \frac{[\{m^4\}\{u\} - \{m^2\}\{m^2u\}]}{[\{1\}\{m^4\} - \{m^2\}^2]}$$

Written out for the case of 5 points, this gives first the table

1	m	m^2	m^3	m^4
1	-2	4	-8	16
1	-1	1	-1	1
1	0	0	0	0
1	1	1	1	1
1	2	4	8	16
Sums 5	0	10	0	34

Hence we have

$$\bar{u}_0 = A = \frac{\left[34 \sum_{m=-2}^{m=2} u_m - 10 \sum_{m=-2}^{m=2} m^2 u_m \right]}{(5)(34) - (10)(10)}$$

$$= \frac{-3u_{-2} + 12u_{-1} + 17u_0 + 12u_1 - 3u_2}{35}$$

Here the smoothed value is a weighted average of the five input values. For the smoothed point $\bar{u}_n$, we have derived the formula

$$\bar{u}_n = \frac{-3u_{n-2} + 12u_{n-1} + 17u_n + 12u_{n+1} - 3u_{n+2}}{35} \qquad (3.3\text{-}1)$$

To analyze this formula we assume, as before, that the signal contains a single frequency and that the sample points are $t_m = m$. Thus we put into formula (3.3-1)

$$u_m = e^{i\omega m}$$

The formula for the transfer function becomes [using (2.4-6)]

$$\frac{17 + 24 \cos \omega - 6 \cos 2\omega}{35} \qquad (3.3\text{-}2)$$

and is, of course, our $H(\omega)$. The transfer functions $H(\omega)$ of these curves are shown in Figure 3.3-1 for 5, 7, 9, and 11 points. The coefficients c_k are, respectively,

$$\tfrac{1}{35}[-3, 12, 17, 12, -3]$$

$$\tfrac{1}{21}[-2, 3, 6, 7, 6, 3, -2]$$

$$\tfrac{1}{231}[-21, 14, 39, 54, 59, 54, 39, 14, -21]$$

$$\tfrac{1}{429}[-36, 9, 44, 69, 84, 89, 84, 69, 44, 9, -36] \qquad (3.3\text{-}3)$$

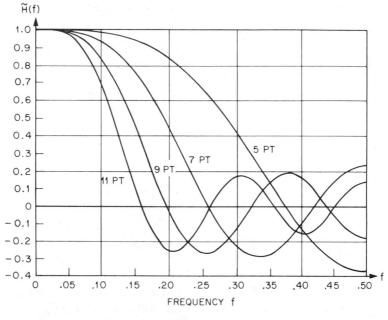

FIGURE 3.3-1 TRANSFER FUNCTION FOR SMOOTHING BY LEAST-SQUARES QUADRATICS

Remember that in the transfer function the noncentral coefficients must be doubled in order to get the corresponding coefficients of the cosines. In Figure 3.3-1 we again plot the independent variable in terms of the rotational frequency f.

These curves of $\tilde{H}(f)$ describe the transfer function of least-squares smoothing by quadratics and resemble those for smoothing by straight lines *except* for the apparent higher degree of tangency of $\omega = 0$. In both cases, using more terms in the smoothing formula causes the curves to come down more rapidly and to be slightly lower in the subsequent wiggles.

We now go to the next case and smooth by fitting least-squares quartics (quintics give the same results). We therefore assume a quartic of the form

$$u(t) = A + Bt + Ct^2 + Dt^3 + Et^4$$

Then using the data points $t = -m, -(m - 1), \ldots, 0, 1, \ldots, m$ (which temporarily fixes the coordinate system and makes the notation much simpler), we set up the differences (the residuals) between the data and the calculated values, square them, and sum over all the data. The next step is to seek the minimum of this function of the variables A, B, C, D, E. As usual, the function to be minimized is differentiated in turn with respect to each of the variables, and the corresponding derivatives are set equal to zero. The result is the set of normal equations. Only the first, third, and fifth of these equations are necessary, since all we need is the value of A. For $m = 3, 4, 5, 6$ or, what is the same thing, for 7, 9, 11, 13 points, we get the value of A, *which will be linear in the original data* u_m. The smoothing formulas obtained have the following coefficients c_k:

$$\tfrac{1}{231}[5, -30, 75, 131, 75, -30, 5]$$

$$\tfrac{1}{429}[15, -55, 30, 135, 179, 135, 30, -55, 15]$$

$$\tfrac{1}{429}[18, -45, -10, 60, 120, 143, 120, 60, -10, -45, 18]$$

$$\tfrac{1}{2431}[110, -198, -135, 110, 390, 600, 677, 600, 390, 110, -135, -198, 110]$$

$$(3.3-4)$$

As usual, we assume the input function $u(t) = e^{i\omega t}$ and obtain the transfer functions $H(\omega)$. These four cases are shown in Figure 3.3-2, where we again use angular frequency f in plotting.

Once more the effect of the higher-degree polynomial is a higher degree of tangency at $\omega = 0$; in addition, the use of extra terms in the smoothing formula makes the curve come down sooner.

This higher degree of tangency at $\omega = 0$ is a general result.

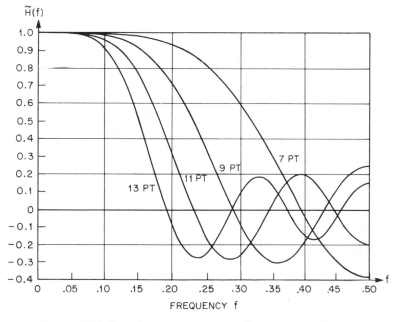

FIGURE 3.3-2 TRANSFER FUNCTION FOR SMOOTHING BY LEAST-SQUARES QUARTICS

THEOREM: *The more powers of* t *that we fit in the time domain, the higher the tangency at* $\omega = 0$ *in the frequency domain.*

To prove this theorem, we assume a smoothing formula of the form

$$y_n = \sum_{k=-N}^{k=N} c_k u_{n-k}$$

Assume also that the coefficients c_k have been determined to make the formula true for $u(t) = 1, t, t^2, \ldots, t^{(m-1)}$ but not true for t^m. This means that the equations (using $n = 0$)

$$\sum_{k=-N}^{k=N} k^p c_k = \begin{cases} 1, & p = 0 \\ 0, & p = 1, \ldots, m-1 \\ \neq 0, & p = m \end{cases}$$

Now consider the transfer function in the form

$$H(\omega) = \sum_{k=-N}^{k=N} c_k e^{-ik\omega}$$

Since the equation is true for $u(t) = 1$, it follows that the preceding equation is true for $\omega = 0$, and therefore $H(0) = 1$. By differentiating with respect to ω and then setting $\omega = 0$, we get the equation (factor out an i)

$$\sum_{k=-N}^{k=N} k c_k = 0$$

From this it follows that $H'(0) = 0$. Repeated differentiation and setting $\omega = 0$ will show that the successive derivatives are zero, up through the $(m-1)$st derivative. But for the mth derivative there will not be cancellation of the terms, and so that corresponding derivative of $H(\omega)$ will not be zero. Thus we have proved the theorem.

Exercises

3.3-1 Derive in detail the transfer function for the 5-point least-squares quartic. Give the answer in terms of cosines.

3.3-2 Derive the transfer function for the 7-point quartic. Give the answer in terms of cosines.

3.3-3 For the formula of exercise 3.3-2, show that the theorem is satisfied.

3.4 MODIFIED LEAST SQUARES

Least squares is a very popular method for approximation. Its chief advantage is that it is easily understood (superficially) and easily computed. The purpose of this section is to point out some of its many faults and to suggest a method for compensating for some of them. As a result, this approach uses least squares to find a first approximation and then modifies this approximation to obtain a more satisfactory result.

In Section 3.2 we fitted local least-squares straight lines to $2N + 1$ equally spaced consecutive data points and obtained the smoothing formula

$$y_n = \frac{1}{2N + 1} \sum_k u_{n-k}$$

We may picture this as the window shown in Figure 3.4-1. We will later see why it is often wise to reduce the two end values to one-half their assigned values. If we do this, the window will look like Figure 3.4-2. The change in the sum of the weights c_k suggests that we also need to change the divisor from $2N + 1$ to $2N$.

The transfer function that we found before was (3.2-1)

$$H(\omega) = \frac{\sin(N + \tfrac{1}{2})\omega}{(2N + 1)\sin(\omega/2)}$$

We need, therefore, to subtract one half of the two end terms

$$\tfrac{1}{2}e^{-i\omega N} + \tfrac{1}{2}e^{i\omega N} = \cos \omega N$$

to get the new formula. After we also adjust the divisor of the transfer function from $2N + 1$ to $2N$ (which is the sum of the coefficients), we have

$$\frac{1}{2N} \left\{ \frac{\sin(N + \tfrac{1}{2})\omega}{\sin(\omega/2)} - \cos \omega N \right\}$$

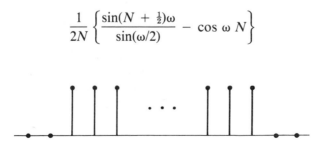

FIGURE 3.4-1 SMOOTHING WINDOW FOR $2N + 1$ POINTS

FIGURE 3.4-2 MODIFIED SMOOTHING WINDOW FOR $2N + 1$ POINTS

Expand the first sine term (and get a common denominator):

$$\frac{[\sin N\omega \cos(\omega/2) + \cos N\omega \sin(\omega/2) - \cos \omega N \sin(\omega/2)]}{2N \sin(\omega/2)}$$

$$= \left\{ \frac{\sin N\omega}{2N \sin(\omega/2)} \right\} \cos(\omega/2) \qquad (3.4\text{-}1)$$

The brackets are the same as before *except* with $2N$ in place of $2N + 1$. This means that the zeros of the transfer function are spaced slightly farther apart, at π/N instead of at $\pi/(N + \frac{1}{2})$. This in turn means that the main lobe is slightly wider. The change of the divisor from $2N + 1$ to $2N$ also means that the term in the brackets would end at a slightly higher value. But the extra $\cos \omega/2$ factor means that the curve will come down somewhat more rapidly. See Figure 3.4-3 and compare with Figure 3.2-2.

If you believe that smoothing is the removal of high frequencies in the signal, then clearly the modified window does a much better job than did the original window. True, the price is indeed a slightly wider main lobe,

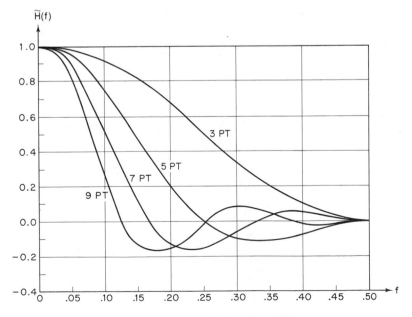

FIGURE 3.4-3 MODIFIED STRAIGHT LINE WINDOWS

now from $-\pi/N$ to π/N, but the gain in the tails of the transfer function seems to be well worth it. Thus modifying the original least-squares filter appears to be a good thing to do. There are, of course, many other ways we might have modified the original least-squares filter to get a more desirable filter.

In a similar fashion, we can modify the quadratic filters of Section 3.3. We merely choose the end constants as a parameter and require that it be fixed so that at the end of the interval, at $f = \frac{1}{2}$, the transfer function shall be zero. This amounts to assuming that the sum of the coefficients when alternating signs are applied totals zero. We have then to adjust the divisor of the filter so that at $f = 0$ the transfer function has the value 1. Corresponding to the earlier formulas (3.3-3), we get the set

$$\frac{1}{96}[7, 24, 34, 24, 7]$$

$$\frac{1}{52}[1, 6, 12, 14, 12, 6, 1]$$

$$\frac{1}{548}[-1, 28, 78, 108, 118, 108, 78, 28, -1]$$

$$\frac{1}{980}[-11, 18, 88, 138, 168, 178, 168, 138, 88, 18, -11] \qquad (3.4\text{-}2)$$

The corresponding transfer functions are shown in Figure 3.4-4. Again we see the slight widening of the main lobe and the drastic reduction in the tails near $f = \frac{1}{2}$.

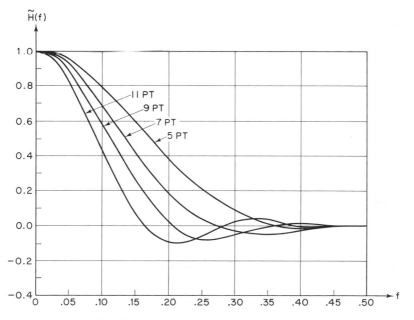

FIGURE 3.4-4 MODIFIED LEAST-SQUARES QUADRATIC SMOOTHING

This is only one way of modifying the least-squares solution to get a more satisfactory smoothing formula. There is no single best method for modifying the original least-squares solution that meets all situations, but it should be clear that modifying a least-squares solution is something to think about in most applications of smoothing.

Exercises

3.4-1   Modify the first of the quartic filters (3.3-4) and sketch its transfer function.

3.4-2   A study of the shape of the array of filter coefficients suggests that

$$c_{2N-k} = C(2N, k)/2^{2N}$$

might be worth examining. Show that the transfer function is $\cos^N \omega$. Plot the first four cases.

3.4-3   Modify the second quartic filter (3.3-4).

3.4-4   Give details of how to modify all the formulas in (3.3-4).

3.5   DIFFERENCES AND DERIVATIVES

The difference operator

$$\Delta u_n = u_{n+1} - u_n$$

provides another class of examples showing the advantages of the frequency approach to a polynomial-derived formula. The repeated use of this operator Δ leads to

$$\Delta^k[u_n] = \Delta[\Delta^{(k-1)}][u_n]$$

Its importance arises from the theorem that the operator Δ^{n+1} annihilates a polynomial $P_n(x)$ of degree n in x; that is,

$$\Delta^{(n+1)}P_n(x) = 0$$

Furthermore, the difference operator can be shown to "amplify" small errors [*H*, Chapters 10 and 35]. Thus the difference table of a function that is identically zero but that has a single error (because of linearity, we can take it to be 1) will have the binomial coefficients with alternating signs in the successive columns of differences. See Table 3.5-1.

TABLE 3.5-1 Difference table

n	$f(n)$	$\Delta f(n)$	$\Delta^2 f(n)$	$\Delta^3 f(n)$
-3	0			
		0		
-2	0		0	
		0		1
-1	0		1	
		1		-3
0	1		-2	
		-1		3
1	0		1	
		0		-1
2	0		0	
		0		0
3	0		0	
		0		
4	0			

Using the frequency approach, we now examine the effect that the operator Δ has on an arbitrary frequency $e^{i\omega t}$; we have

$$\Delta e^{i\omega t} = e^{i\omega(t+1)} - e^{i\omega t}$$

$$= [e^{i\omega} - 1]e^{i\omega t}$$

$$= e^{i\omega/2}[e^{i\omega/2} - e^{-i\omega/2}]e^{i\omega t}$$

$$= ie^{i\omega/2}\left[2 \sin \frac{\omega}{2}\right]e^{i\omega t}$$

It follows that k repeated applications of the Δ operator will give

$$\Delta^k[e^{i\omega t}] = [i^k]e^{ik\omega/2}\left[2 \sin \frac{\omega}{2}\right]^k e^{i\omega t}$$

The first two factors are of size 1, and so the *amplification* at the frequency ω is contained in the factor

$$\left[2 \sin \frac{\omega}{2}\right]^k$$

where $-\pi \leq \omega \leq \pi$ is the usual Nyquist interval. We see immediately that for the smallest one-third of the frequencies, $0 \leq \omega < \pi/3$, the difference operator Δ decreases the amplitude of any frequency, whereas in the upper two thirds of the frequencies, $\pi/3 < \omega \leq \pi$, there is amplification. See Figure 3.5-1. This situation explains the typical use of the difference table to locate

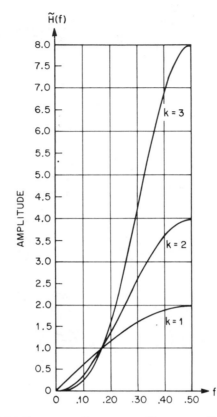

FIGURE 3.5-1 FREQUENCY RESPONSE OF DIFFERENCE OPERATOR Δ^k

(high-frequency) noise. Noise in this case means "the upper two-thirds of the Nyquist interval." We will later show how to filter other intervals than the $-\frac{1}{3}$ to $\frac{1}{3}$, as well as how to do it in a "minimum-maximum error" (Chebyshev sense).

Differences are also used to approximate derivatives. For instance, the central difference formula

$$u_{n+1} - u_{n-1} = 2hu_n'$$

provides an approximation to the derivative. Using our frequency approach, we set

$$u(t) = e^{i\omega t}$$

The true derivative is

$$u'(t) = i\omega e^{i\omega t}$$

From the formula (with $h = 1$) the calculated value is

$$e^{i\omega} - e^{-i\omega} = 2i \sin \omega$$

The ratio of the calculated to the true answer (which is $i\omega$) is

$$R = \frac{\text{calculated}}{\text{true}} = \frac{2i \sin(\omega)}{2i\omega} = \frac{\sin \omega}{\omega}$$

Thus when $\omega = 0$ (dc), the value of the ratio is 1, as it should be, but for all other ω the ratio is less than 1. Thus the formula underestimates the value of the derivative for all other frequencies in the Nyquist interval.

For the second derivative, we use the estimate

$$u''(t) = u(t + 1) - 2u(t) + u(t - 1)$$

and obtain the ratio of the calculated to true answers as

$$R = \frac{\text{calculated}}{\text{true}} = \frac{[e^{i\omega} - 2 + e^{-i\omega}]}{(i\omega)^2}$$

$$= \frac{2(1 - \cos \omega)}{\omega^2}$$

$$= \left[\frac{\sin(\omega/2)}{\omega/2}\right]^2$$

which is the square of the preceding curve but contracted by a factor of 2 in the independent variable.

The beginner should take some particular frequency and see how these formulas agree with practice by operating on definite numbers and, in particular, notice how the zero estimates arise.

Exercises

3.5-1 Apply the theory of the preceding section to the following third derivative approximation:

$$[-u(n + 2) + 2u(n + 1) - 2u(n - 1) + u(n - 2)]/2$$

3.5-2 Discuss the tangency of the transfer function of the kth derivative at $\omega = 0$.

3.5-3 Prove by induction that $\Delta^{m+1} P_m(x) = 0$.

3.5-4 Plot the transfer functions for the two estimates of derivatives.

3.6 MORE ON SMOOTHING: DECIBELS

Sections 3.2 and 3.3 show that many smoothing formulas using least squares keep the value at zero frequency (dc) correct but, in general, decrease the amount of any higher frequency that may be in the function being smoothed. The curves in the figures are the *transfer functions of the linear process of smoothing*; that is, they are, for each ω, the corresponding eigenvalue of the process in the range up to the Nyquist frequency (which is where aliasing begins).

This situation suggests examining other classical smoothing formulas in order to see if they have a similar property. Perhaps the two best-known smoothing formulas that have come down to us from the past are Spencer's 15- and 21-point smoothing formulas [KS3, p. 372]

$$y_n = \tfrac{1}{320}[-3u_{n-7} - 6u_{n-6} - 5u_{n-5} + 3u_{n-4} + 21u_{n-3}$$
$$+ 46u_{n-2} + 67u_{n-1} + 74u_n + 67u_{n+1} + 46u_{n+2}$$
$$+ 21u_{n+3} + 3u_{n+4} - 5u_{n+5} - 6u_{n+6} - 3u_{n+7}] \qquad (3.6\text{-}1)$$

and

$$y_n = \tfrac{1}{350}[-u_{n-10} - 3u_{n-9} - 5u_{n-8} - 5u_{n-7} - 2u_{n-6}$$
$$+ 6u_{n-5} + 18u_{n-4} + 33u_{n-3} + 47u_{n-2} + 57u_{n-1}$$
$$+ 60\,u_n + 57u_{n+1} + \cdots] \qquad (3.6\text{-}2)$$

The transfer functions for these two formulas are given in Figure 3.6-1. If we accept the hypothesis that they were also designed to remove noise, then we are forced to the conclusion that noise was classically identified with high frequencies and message or information with low frequencies. The two formulas differ in how much they let pass through the filtering (smoothing) process. The longer formulas has the shorter passband of frequencies and is about what would be expected from general experience in science.

These formulas were designed not only to pass low frequencies and stop high frequencies but also to be easy to compute by hand. Consequently, they are not necessarily optimal for present computers. Indeed, for the number of terms and the given "cutoff frequency," better formulas can easily be designed (see Chapters 6 and 9).

As shown in Figure 3.6-1, the curves are not too informative, since they are so small at the higher frequencies that we cannot see how good they are. It is therefore better to plot the logs of the numbers $|H(\omega)|$. For this purpose, it is customary to use decibels (tenths of a bel), abbreviated dB, and defined as

$$20 \log |\text{ratio}| = \text{decibel units} \qquad (3.6\text{-}3)$$

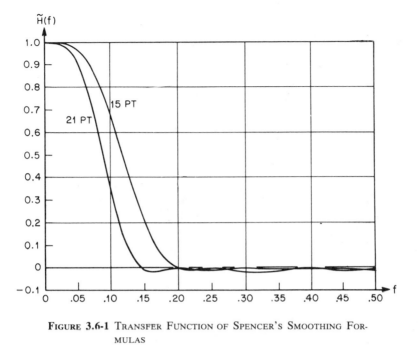

FIGURE 3.6-1 TRANSFER FUNCTION OF SPENCER'S SMOOTHING FORMULAS

where, of course, we pick the reference value (the denominator of the ratio) as the original amount of the frequency present in the signal. Figure 3.6-2 shows this version of the Spencer's smoothing formulas. The past inference that noise was high frequency and signal low frequency is strengthened as we examine this new plot of the corresponding transfer functions.

Since

$$20 \text{ dB} = \text{factor of } 10$$

we have the following table:

dB	Factor
20	10
40	100
60	1000
80	10^4
100	10^5

It is now clear that smoothing formulas generally remove some frequencies and let others pass. In 1927 Slutsky and Yule reported on this statistical effect [KS3, p. 378]: that smoothing could greatly affect what was later found by further analysis, especially when examining data with large amounts of noise. As a result of this filtering effect, sometimes the periods

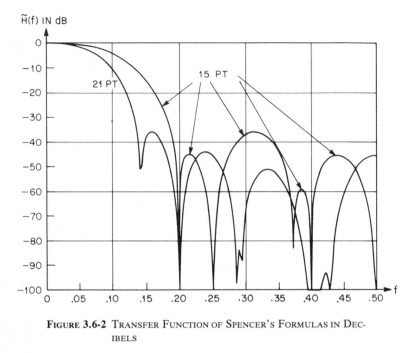

$\tilde{H}(f)$ IN dB

FIGURE 3.6-2 TRANSFER FUNCTION OF SPENCER'S FORMULAS IN DEC-
IBELS

that are found in smoothed data are more the effect of the smoothing than
of the original data (see Chapter 10).

3.7 MISSING DATA AND INTERPOLATION

In a long record of data there are often one or more (isolated) missing
values. This situation occurs for various reasons: the measurements may
never have been made; they may have been misrecorded and thus later
removed; or the formula used to compute the successive values of the function may have involved an indeterminate form, such as $(\sin x)/x$ at $x = 0$,
and the computer refused to divide by zero.

Perhaps the most common way of supplying an isolated missing value
is to use an interpolation formula based on the assumption that the data
locally is a polynomial of some odd degree. This is equivalent to the assumption that the next higher-order difference is zero. For instance, assume
that the fourth difference is zero. We therefore have

$$\Delta^4 u_{n-2} \equiv u_{n-2} - 4u_{n-1} + 6u_n - 4u_{n+1} + u_{n+2} = 0$$

Solving this equation for u_n gives the very convenient, stable formula for the missing value

$$u_n - \tfrac{1}{6}[-u_{n-2} + 4u_{n-1} + 4u_{n+1} - u_{n+2}] \qquad (3.7\text{-}1)$$

The noise amplification factor of this formula (Section 1.7) is 17/18.

To compute the noise amplification in the general case, we need to know that the sum of the squares of the binomial coefficients of order N is $C(2N, N)$. Hence for the $2k$th difference formula, we have the noise amplification

$$\frac{C(4k, 2k) - C^2(2k, k)}{C^2(2k, k)} \approx \sqrt{\frac{\pi}{2}k} - 1 \qquad (3.7\text{-}2)$$

For $k = 1$ this formula gives $\tfrac{1}{2}$; for $k = 2$, $\tfrac{17}{18}$; for $k = 3$, 1.31; and so on.

Notice that supplying one isolated missing value is not the same as the usual dynamic process of imagining a stream of data passing through a digital filter. Nevertheless, we can look at what this formula does to any given frequency and thus look at its transfer function. To do so, we proceed as usual and substitute for the function u_n the complex frequency $e^{i\omega n}$. We find for equation (3.7-1)

$$H(\omega) = \tfrac{1}{3}[4 \cos \omega - \cos 2\omega]$$

as the transfer function, which should give the value 1 if the interpolated value is exactly what it should be. It is easy to see that we obtain the right answer for frequency 0. For higher frequencies, the value is not too good, particularly for very high frequencies. If the result seems to be strange, as shown in Figure 3.7-1, then trying the limiting function in the Nyquist interval, that is, $1, -1, 1, -1, \ldots,$ will reveal why the curves are what they are. Negative values on the graph of the transfer function imply a change in sign. The figure also shows the transfer function for sixth- and eighth-order differences equated to zero.

Again we see that the new way of looking at an old technique shows much more about what the formula does than is evident from mere inspection. We see the power of the frequency approach via the eigenfunctions of a linear, equally spaced sampled system. These formulas clearly show the danger of interpolating a missing value when the data is noisy, which means that the data has numerous high frequencies.

Similar curves can be drawn for formulas that use the least-squares polynomial fit to the adjacent values as the basis for interpolating the missing value.

If we now turn to the problem of interpolating midpoint values in some

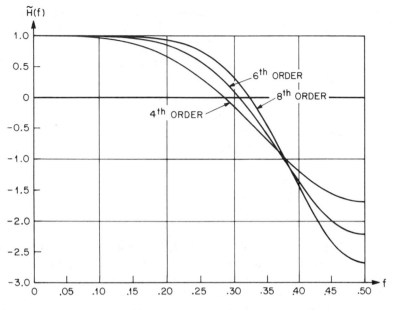

FIGURE 3.7-1 TRANSFER FUNCTION FOR MISSING DATA

data, the simplest (linear) interpolation formula is

$$\tfrac{1}{2}(u_{n+1/2} + u_{n-1/2})$$

This leads to the transfer function $\cos \omega/2$. Using four adjacent points and making the formula exact for all cubics gives the formula

$$\tfrac{1}{16}(-u_{n+3/2} + 9u_{n+1/2} + 9u_{n-1/2} - u_{n-3/2})$$

which in turn leads to the transfer function

$$\tfrac{1}{8}(9 \cos \omega/2 - \cos 3 \omega/2)$$

The topic of interpolation will be treated in more detail in Section 7.1, where the transfer functions of these formulas appear.

Exercises

3.7-1 Derive the two midpoint interpolation formulas given in the text. Plot the transfer functions.

3.7-2 In digital signal processing the data is often given at the midpoints and needs to be shifted $\Delta t/2$. We fitted 2 and 4 points and found the interpolatory formula transfer functions for the midpoints. Derive the corresponding linear least-squares formulas for 4 and 6 points. Plot the transfer functions and compare with Figure 7.1-1.

3.7-3 Same as exercise 3.7-2, except use quadratic on 6 points.

3.8 A CLASS OF NONRECURSIVE SMOOTHING FILTERS

Any evaluation method almost always leads in a short time to a corresponding design problem—how to get the desired evaluation. Having looked at a large number of classical smoothing filters from the frequency approach, we are ready to attempt the *design* of a smoothing filter. At this point we will try only the simple class of symmetric filters,

$$y_n = au_{n-2} + bu_{n-1} + cu_n + bu_{n+1} + au_{n+2}$$

As usual, we start by asking what happens to a single frequency when we apply this smoothing filter; that is, we assume that

$$u_n = e^{i\omega n}$$

and examine the output. After putting this function in the equation and factoring out the exponential depending on n, we obtain the transfer function

$$H(\omega) = 2a \cos 2\omega + 2b \cos \omega + c$$

If the symmetry had not been assumed, there would have been some sine terms with imaginary coefficients.

How shall we choose the coefficients a, b, and c in our three-parameter family of filters? The answer depends, of course, on what we want to do, which frequencies we want to pass and which we want to stop. Suppose that, as usual, we want to pass low frequencies and stop high frequencies. Thus we want a low-pass filter. To convert a low-pass filter to a high-pass filter, we need only compute $u_n - y_n$ as the new filter; what was stopped in the y_n is present in the u_n and is passed, whereas what is passed by the y_n is eliminated in the difference $u_n - y_n$.

We start by *arbitrarily* imposing two conditions on the transfer function; at the low-frequency end we require

$$H(0) = 1 \qquad \text{exact at dc (lowest frequency)}$$

and at the upper end we require

$$H(\pi) = 0 \qquad \text{no highest frequency gets through}$$

These two conditions are equivalent to the pair of equations

$$H(0) = 2a + 2b + c = 1$$
$$H(\pi) = 2a - 2b + c = 0$$

From these two equations we get

$$b = \tfrac{1}{4}$$
$$c = \tfrac{1}{2} - 2a$$

and we are reduced to a one-parameter family of filters (see Figure 3.8-1). The filters are

$$H(\omega) = 2a \cos 2\omega + \frac{1}{2} \cos \omega + \frac{1}{2} - 2a$$

$$= 4a \cos^2 \omega - 2a + \frac{1}{2} \cos \omega + \frac{1}{2} - 2a$$

$$= 4a \left[\cos^2 \omega + \frac{\cos \omega}{8a} + \frac{1}{8a} - 1 \right]$$

$$= 4a \left[\cos \omega + 1 \right] \left[\cos \omega + \left(\frac{1}{8a} - 1 \right) \right]$$

Since $H(\pi) = 0$ the factor $[\cos \omega + 1]$ had to occur. Note that $H(\omega)$ is a periodic function of ω and that it is an even function; however, as noted, because of sampling and the resulting aliasing, there is little meaning to the transfer function outside the Nyquist interval. The periodicity of the mathematical expression reflects the aliasing rather than any analytic property.

From Figure 3.8-1 we can select a filter that approximately meets our needs. Or we could impose one more condition on the transfer function $H(\omega)$ and thus directly determine the filter. We will try this second approach in order to illustrate the designing of very simple filters.

As a first example of the direct design of a filter, suppose that we require

$$H_1 \left(\frac{\pi}{2} \right) = 1$$

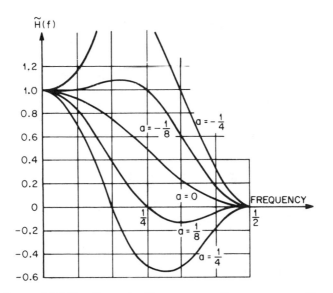

FIGURE 3.8-1 The Transfer Function as a Function of the Parameter

which is equivalent to

$$H_1\left(\frac{\pi}{2}\right) = -2a + \frac{1}{2} - 2a = 1$$

This gives

$$a = -\tfrac{1}{8}, \qquad c = \tfrac{3}{4}$$

Therefore, the filter is

$$y_n = \tfrac{1}{8}[-u_{n-2} + 2u_{n-1} + 6u_n + 2u_{n+1} - u_{n+2}]$$

and its transfer function is

$$H_1(\omega) = -\frac{\cos 2\omega}{4} + \frac{\cos \omega}{2} + \frac{3}{4}$$

$$= -\frac{1}{2}[\cos \omega + 1][\cos \omega - 2]$$

As a second example of the design of a filter, suppose that we wish to balance the two halves of the filter by requiring

$$H_2\left(\frac{\pi}{2}\right) = \frac{1}{2}$$

Then we get, in turn,

$$H_2\left(\frac{\pi}{2}\right) = -2a + \frac{1}{2} - 2a = \frac{1}{2}$$

$$a = 0, \qquad c = \frac{1}{2}$$

Hence

$$y_n = \tfrac{1}{4}[u_{n-1} + 2u_n + u_{n+1}]$$

and

$$H_2 = \tfrac{1}{2} \cos \omega + \tfrac{1}{2}$$

As a third example, suppose that we try to do as well as possible in the neighborhood of zero frequency. We already have both

$$H_3(0) = 1 \quad \text{and} \quad \frac{dH_3(0)}{d\omega} = 0$$

So we impose the further condition that

$$\frac{d^2 H_3(0)}{d\omega^2} = 0 = -8a - \frac{1}{2}$$

or

$$a = -\tfrac{1}{16}$$

Thus we have

$$y_n = \tfrac{1}{16}[-u_{n-2} + 4u_{n-1} + 10u_n + 4u_{n+1} - u_{n+2}]$$

with

$$H_3(\omega) = \frac{\cos 2\omega}{8} + \frac{\cos \omega}{2} + \frac{5}{8}$$

$$= -\frac{1}{4}[\cos \omega + 1][\cos \omega - 3]$$

These filters are merely some of the simplest ones that we can design and are mainly for illustrative purposes. They are not intended as serious design problems.

3.9 AN EXAMPLE OF HOW A FILTER WORKS

Let us now examine *how* a filter does its job. For this purpose, we want the simplest example that we can get. We first select two frequencies in the Nyquist interval, one at $f = \frac{1}{6}$ and the other at $f = \frac{1}{3}$. Next, we design the simple filter to exactly pass the first and stop the second; that is, we pick our filter form as

$$y_n = au_{n+1} + bu_n + au_{n-1}$$

The transfer function is, therefore,

$$H(f) = 2a \cos 2\pi f + b$$

where we are now working in the rotation variable f instead of the earlier angular frequency variable ω. The two conditions on the filter that we are imposing are

$$\tilde{H}(\tfrac{1}{6}) = 1 \quad \text{and} \quad \tilde{H}(\tfrac{2}{6}) = 0$$

We get the two equations

$$\tilde{H}(\tfrac{1}{6}) = 1 = 2a(\tfrac{1}{2}) + b$$
$$\tilde{H}(\tfrac{2}{6}) = 0 = 2a(-\tfrac{1}{2}) + b$$

Therefore

$$b = \tfrac{1}{2}, \qquad a = \tfrac{1}{2}$$

and the filter is

$$y_n = \tfrac{1}{2}[u_{n-1} + u_n + u_{n+1}]$$

Writing just the coefficients we have

$$\tfrac{1}{2}[1, 1, 1]$$

We now need to make some input data to test the filter to see if it does as we expect it to do. What do we mean by data of frequency $f = \frac{1}{6}$? We mean exactly the sequence of numbers given by

$$\cos[2\pi(\tfrac{1}{6})n] = \cos\frac{\pi n}{3}$$

For

$$n = 0 \qquad \cos\frac{\pi 0}{3} = 1$$

$$n = 1 \qquad \cos\frac{\pi}{3} = \frac{1}{2}$$

$$n = 2 \qquad \cos\frac{2\pi}{3} = -\frac{1}{2}$$

$$n = 3 \qquad \cos\pi = -1$$

etc.

We see these numbers in column B of Table 3.9-1. In the table are the input data for frequencies $0, \frac{1}{6}, \frac{2}{6}, \frac{3}{6}$ in the columns headed A, B, C, and D.

These cases can be tested by putting the coefficients of the filter opposite any three consecutive values in a column and noting what comes out of the filter. In column A you should get $\frac{3}{2}$ every time as the transfer function $\tilde{H}(f)$ (the eigenvalue) of the filter is $\frac{3}{2}$. For column B you should get exactly the middle value because the eigenvalue is 1. For column C you should get 0 always. Finally for column D you should get $-\frac{1}{2}$ times the middle value of the three. Make sure you try these out to see that what was said is realized, that a digital filter actually does what it is supposed to do.

This shows that the pure frequencies behave as they should. We next test sums of inputs and show that the filter acts on each frequency as it

TABLE 3.9-1 Data to illustrate filtering

n	A $\tilde{H}(0) = \frac{3}{2}$	B $\tilde{H}(\frac{1}{6}) = 1$	C $\tilde{H}(\frac{2}{6}) = 0$	D $\tilde{H}(\frac{1}{2}) = -\frac{1}{2}$
0	1	1	1	1
1	1	$\frac{1}{2}$	$-\frac{1}{2}$	-1
2	1	$-\frac{1}{2}$	$-\frac{1}{2}$	1
3	1	-1	1	-1
4	1	$-\frac{1}{2}$	$-\frac{1}{2}$	1
5	1	$\frac{1}{2}$	$-\frac{1}{2}$	-1
6	1	1	1	1
7	1	$\frac{1}{2}$	$-\frac{1}{2}$	-1
8	1	$-\frac{1}{2}$	$-\frac{1}{2}$	1

TABLE 3.9-2 More data to illustrate filtering

n	B + C → B	A + B + C → $\frac{3}{2}$ + B	B + C + D → B $\oplus \frac{1}{2}$ D
0	2	3	3
1	0	1	−1
2	−1	0	0
3	0	1	−1
4	−1	0	0
5	0	1	−1
6	2	3	3
7	0	1	−1
8	−1	0	0

should without regard to any other frequency present. Thus in Table 3.9-2 we have added the indicated columns and shown what we expect to come out from filtering them with the given filter.

This is the simplest filter for which we can show how a filter actually works. You should study the tables and experiment for yourself to see that each frequency is multiplied by its transfer function value (eigenvalue) and occurs again in the sum with its proper multiplier. Thus you see that a digital filter:

1. decomposes the input into its separate frequencies

2. multiplies each by its proper transfer function value (eigenvalue)

3. adds them all together again to form the output

Be sure to see this clearly; it is the heart of digital filtering! Each eigen-function (input) gets multiplied by its eigenvalue independently of what other eigenfunctions are present, and the appropriate sum is put out.

Exercises

3.9-1   For the filter in this section use $f = \frac{1}{4}$, find the eigenvalue ($\tilde{H}(\frac{1}{4})$) and check the corresponding data.

3.9-2   Design a filter for $\tilde{H}(\frac{1}{8}) = 1$, $\tilde{H}(\frac{3}{8}) = 0$. Using $1/\sqrt{2} = 0.7$ and $(0.7)^2 = \frac{1}{2}$ check these two frequencies. *Answer:* $\frac{1}{2}[0.7, 1.0, 0.7]$

3.9-3   For the previous problem, check the sum of the two frequencies as well as $f = 0$ and $f = \frac{1}{2}$.

3.9-4   Design a corresponding filter for $\tilde{H}(0) = 1$ $\tilde{H}(\frac{1}{2}) = 0$ and test at $\tilde{H}(\frac{1}{4})$.

3.9-5   Design a filter for $\tilde{H}(0) = 0$, $\tilde{H}(\frac{1}{2}) = 1$ and test for $\tilde{H}(\frac{1}{4})$.

3.9-6 Design a filter using $\bar{H}(0) = 1$, $\bar{H}(\frac{1}{4}) = 0$. Test it on suitable values such as $f = \frac{1}{8}, \frac{3}{8}, \frac{1}{2}$, as well as the given ones.

3.9-7 Design and test a filter for which $\bar{H}(\frac{1}{8}) = 0$, $\bar{H}(\frac{3}{8}) = 1$. (See Exercise 3.9-2.)

3.10 INTEGRATION: RECURSIVE FILTERS

Another well-known operation that uses a linear combination of the data is *numerical integration*; in particular, the reader is probably familiar with the trapezoid rule, the midpoint rule, and Simpson's formula. Let us examine these formulas from the frequency point of view.

The trapezoid rule is (using $y_0 = 0$)

$$y_{n+1} = y_n + \tfrac{1}{2}[u_{n+1} + u_n]$$

where the u_n are the integrand values and the y_n are the integral values (area values). Since y occurs on both sides of this equation, this filter is clearly recursive (Section 1.1). On the other hand, the earlier smoothing formulas considered were nonrecursive filters. In this case, we assume that the input $u(t)$ is $e^{i\omega t}$ and, since the equation is linear, that the corresponding output $y(t)$ is of the form $A(\omega)e^{i\omega t}$ (Section 2.5). Solving for $A(\omega)$, we obtain

$$A(\omega) = \frac{1}{2}\frac{[e^{i\omega} + 1]}{[e^{i\omega} - 1]}$$

Divide numerator and denominator by $e^{i\omega/2}$ and write the result in terms of the conventional trigonometric functions:

$$A(\omega) = \frac{\cos \omega/2}{2i \sin \omega/2}$$

The true answer for integration of the function $e^{i\omega t}$ is, of course, $[1/i\omega]e^{i\omega t}$. We now take the ratio of the calculated to true:

$$\frac{\text{calculated}}{\text{true}} = \cos\frac{\omega}{2}\left[\frac{(\omega/2)}{\sin(\omega/2)}\right]$$

At $\omega = 0$, this ratio is clearly 1 and falls to zero at the Nyquist frequency, $\omega = \pi$ ($f = \frac{1}{2}$), which is the natural boundary, for it is there that aliasing begins. See Figure 3.10-1.

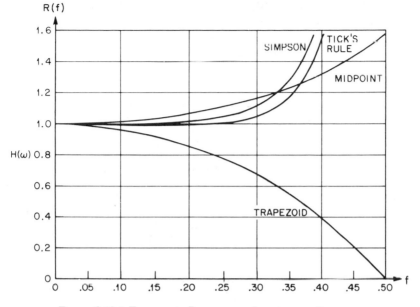

FIGURE 3.10-1 FREQUENCY RESPONSE OF INTEGRATION FORMULAS

Near $f = 0$ the curves are hard to read so we append a table of values (Table 3.9-3).

Why have we taken the ratio of the computed to true answers instead of simply evaluating the transfer function as in the past? The answer is simple; for both smoothing and filtering, we took the ratio of the output to the input, which was the natural comparison to see how well we did. But in operations like differentiation and integration we compare the result obtained with the result wanted, and so the ratio of the calculated to true values is used for evaluating such formulas. Clearly, the right side is a *transfer function* in the sense that multiplying by it transfers us from the true values to the calculated values as a function of ω.

To get a better feeling for the transfer functions, we expand (for the

TABLE 3.9-3 Values of transfer functions

f	Trapezoid	Simpson	Tick's rule
0.06	-0.0104	0.0001	-0.0015
0.08	-0.0185	0.0003	-0.0025
0.10	-0.0292	0.0008	-0.0037
0.12	-0.0426	0.0017	-0.0048
0.14	-0.0587	0.0032	-0.0059
0.16	-0.0778	0.0056	-0.0065
0.18	-0.1002	0.0092	-0.0064
0.20	-0.2261	0.0146	-0.0051

trapezoid rule) both numerator and denominator in power series and then divide them out to get

$$R(\omega) = \frac{\text{calculated}}{\text{true}} = 1 - \frac{\omega^2}{12} + \frac{\omega^4}{720} + \cdots$$

which gives the quadratic shape of the ratio near $\omega = 0$.

This same process on Simpson's formula for integration yields for the formula

$$y_{n+1} = y_{n-1} + \tfrac{1}{3}(u_{n+1} + 4u_n + u_{n-1})$$

the transfer function

$$H(\omega) = \frac{e^{i\omega} + 4 + e^{-i\omega}}{3(e^{i\omega} - e^{-i\omega})}$$

Therefore, the ratio of calculated to true is

$$R(\omega) = \frac{\text{calculated}}{\text{true}} = \frac{[2 + \cos \omega]}{[3(\sin \omega)/\omega]} = 1 + \frac{\omega^4}{180} + \cdots$$

This curve begins at $\omega = 0$ with the value 1, has (according to the earlier theorem of Section 3.3) a tangency through the third derivative, and rises initially like a quartic until at $\omega = \pi$ the denominator becomes zero. (See Figure 3.10-1.)

Since this behavior at $\omega = \pi$ is surprising, how can we understand it? Clearly, Simpson's formula amplifies the upper part of the Nyquist interval (the higher frequencies), whereas the trapezoid rule damps them out. At the Nyquist frequency ($\omega = \pi$, or $f = \frac{1}{2}$) the function being integrated could have the values $+1, -1, +1, -1, +1, -1, \ldots$, and in Simpson's formula these values are going to be multiplied by the coefficients 1, 4, 2, 4, 2, 4, 2, 4, . . . , all divided by 3, of course. Looking at the products, we see that each pair of values, the second and third (-1×4 and $+1 \times 2$), the fourth and fifth, and so on, combines to give a negative number; thus for this high-frequency function the integrated sum will tend for long intervals to grow in size in a linear fashion. This effect is known in numerical analysis but is seldom mentioned in textbooks.

The corresponding effect for the trapezoid rule will not show this effect. For frequencies near the Nyquist frequency, we see that long stretches of values will not tend to combine as they do for Simpson's integration formula.

Next, we look at the midpoint integration formula (using $y_0 = 0$):

$$y_{n+1} = y_n + u_{n+1/2}$$

The result of substituting $e^{i\omega t}$ for $u(t)$, $A(\omega)e^{i\omega t}$ for $y(t)$, and computing the ratio of the calculated to true values, is

$$\frac{\omega/2}{\sin(\omega/2)}$$

This expression begins at 1 and slowly rises as ω increases. At the Nyquist limit, we have $\pi/2$ as the value of the function.

The fourth curve of Figure 3.10-1 is the transfer function of a formula due to Leo Tick. The formula was designed to have a transfer function that is as close to unity as we can make it throughout the lower half of the Nyquist interval while still involving only three consecutive terms. The formula is (using $y_0 = 0$)

$$y_{n+1} = y_{n-1} + h(0.3584u_{n+1} + 1.2832u_n + 0.3584u_{n-1})$$

We will derive this formula in Section 13.10.

In summary, we have shown how three classical numerical integration formulas look when examined from the frequency point of view. The use of the natural eigenfunctions of the problem $e^{i\omega t}$ has, as expected, shed a new and different light on each of them. The implications of what the formulas do to noisy (high frequency) data was intuitively understood in hand-calculating days but has not, generally speaking, appeared in modern textbooks. It is clear that in the presence of noise, which usually includes a good deal of high frequency, Simpson's formula is more dangerous to use than are the trapezoid or midpoint formulas. But when there is relatively little high frequency in the function being integrated, then the flatness of Simpson's formula for low frequencies shows why it is superior. The choice to make (once shown) depends in an obvious way on the frequency characteristics of the function being integrated.

A word of caution, however. Actually, it is not possible to judge a method of computation without considering what is to be done with the results. If, for instance, the integrated function is to be analyzed for frequencies, then it is particularly important to examine the effects of the method of integration on various frequencies and to make allowances in the interpretation for these effects. As noted, Tick's integration rule of Figure 3.10-1 was designed to be as accurate as it could be (given the basic form) for the lower half of the Nyquist interval, $0 \le f \le \frac{1}{4}$. The figure shows (just barely) that the error at the upper limit of the band in which it was to be accurate is the same as the maximum error at any earlier point except that the sign is different. Thus the maximum error for any frequency in this interval is minimized. This criterion will be studied more closely in Chapter 13. The price of this accuracy over a large range of frequencies is, among other factors, less accuracy near $\omega = 0$.

Exercises

3.10-1   Apply the methods of this section to the three-eighths rule of integration:

$$y_{n+2} = y_{n-1} + \tfrac{3}{8}[u_{n+2} + 3u_{n+1} + 3u_n + u_{n-1}]$$

3.10-2   Compute and plot the transfer functions for the Newton-Cotes integration formulas for degrees = 2, 3, . . . , 10 [H, p. 342].

3.10-3   Compute the sum of the squares of the coefficients of the u_n terms (the noise amplification of Section 1.7) for the four integration formulas of this section.

3.10-4   Write the transfer function for Tick's rule.

3.10-5   Extend the table of the transfer functions for the four formulas in this section, with special attention to the neighborhood of the origin, but also fairly accurately out to $f = \tfrac{1}{2}$. Examine the values at $f = \tfrac{1}{4}$

3.11   SUMMARY

The purpose of this chapter was to show that the frequency approach provides a useful way to understand the various results of classical polynomial approximation. The examples were chosen for their ease of understanding, general familiarity, and usefulness in computing practice rather than for their basic importance.

As noted before, almost any evaluation method leads to a design problem of how to meet a desired evaluation criterion. This is the design problem we now face.

Before proceeding to the design of filters in general, it is necessary to develop some of the formal mathematics related to the Fourier series, which is the subject of the next two chapters. At each stage we shall include simple examples to illustrate the theory as it is developed, instead of going through a long stretch of formal mathematics before showing how and why it is relevant.

4

Fourier Series: Continuous Case

4.1 NEED FOR THE THEORY

The preceding chapter has shown that the transfer function of the typical smoothing filter is the sum of cosines of various integral frequencies. On the other hand, differentiators were sums of sines, and integrators were ratios of sines and cosines. We will for some time concentrate on the nonrecursive filters (smoothers and differentiators) and leave the recursive filters to the later chapters.

In the latter parts of Section 3.7 we dealt with some filters that involved points at the half-odd integers and the corresponding transfer functions have similar frequencies. There is a tendency to neglect such kinds of filters, but we will examine them occasionally as they often are very useful. They are widely used in numerical analysis, but are usually avoided in digital filter texts.

The filters considered were of three types. First, we had filters that ideally were either 0 or 1 at various places in the Nyquist interval: 0 where we wished to stop and 1 where we wanted to pass the frequencies of the signal. A common special case is the noise filter that typically passes the low frequencies and attenuates the high ones, the cutoff place depending on where we think the frequencies of the signal end. Second, we had differentiators where ideally we wished to approximate the function $H(\omega) = i\omega$ and were concerned with the ratio of the calculated to true values rather than simply the output of the filter. Finally, we had integrators where we wanted to approximate the function $1/i\omega$ and again were interested in ratios. Clearly, other types of filters could occur, and we need to be able to cope with any reasonably shaped transfer function.

To design nonrecursive filters, rather than merely evaluate them, we must go fairly directly from any proposed transfer function to the expansion in terms of the trigonometric functions. It will then be easy to go to the actual coefficients c_k of the digital filter. Fundamental to this design process, therefore, is the expansion of an arbitrary function in terms of the trigonometric functions. This is *the theory of Fourier series*, the subject of this and the next chapter. The theory has much wider applications than to the subject of digital filters.

The relationship of formal mathematics to the real world is ambiguous. Apparently, in the early history of mathematics the mathematical abstractions of integers, fractions, points, lines, and planes were fairly directly based on experience in the physical world. However, much of modern mathematics seems to have its sources more in the internal needs of mathematics and in esthetics, rather than in the needs of the physical world. Since we are interested mainly in *using* mathematics, we are obliged in our turn to be ambiguous with respect to mathematical rigor. Those who believe that mathematical rigor justifies the use of mathematics in applications are referred to Lighthill [Li] and Papoulis [P] for rigor; those who believe that it is the usefulness in practice that justifies the mathematics are referred to the rest of this book. We adopt the attitude that useful mathematics can be mathematically justified, even if it is necessary to alter classical definitions and postulates of mathematics (recall the recent change from "function" to "generalized function" to justify the widespread use of the Dirac delta function). Furthermore, since we are interested in the anatomy of the mathematics, we shall ignore many of the mathematically pathological cases. The fact that we are dealing with samples of a physical function implies that we are trying to understand a reasonable situation.

In short, for our purposes, the justification of the various mathematical models and their rigor rests more on their usefulness in the real world than on the internal esthetics of mathematics.

4.2 ORTHOGONALITY

Although we shall want to expand $H(\omega)$ or $\tilde{H}(f)$, where ω and f are, respectively, the independent variable, it is convenient to present the theory of Fourier series in terms of the neutral variable θ. It is necessary to use such an independent variable because we will later apply the resulting formulas in a number of different ways. Furthermore, the use of the variable θ is widespread and fits in with other texts.

The first basic, widely used concept needed is that of *orthogonality*. Two functions $g_1(\theta)$ and $g_2(\theta)$ (neither identically zero) are said to be *or-*

thogonal with respect to a weight function $K(\theta) \geq 0$, in an interval $a \leq \theta \leq b$, if

$$\int_a^b K(\theta)g_1(\theta)g_2(\theta)\, d\theta = 0$$

This concept is a large extension of the idea of orthogonal lines in n-dimensional space. To see this point, consider two n-dimensional vectors $U = \{u_1, u_2, \ldots, u_n\}$ and $V = \{v_1, v_2, \ldots, v_n\}$. In forming the sum of the two vectors, we add components term by term. The sum is the third side of the triangle with U and V being the other two sides. If it is to be a right triangle with U and V as the legs, we apply the Pythagorean theorem and assert that

$$(U + V)^2 = U^2 + V^2$$

In terms of the components, this is

$$\sum_{i=1}^n (u_i + v_i)^2 = \sum_{i=1}^n u_i^2 + \sum_{i=1}^n v_i^2$$

Multiply this out and cancel the square terms from both sides, leaving the sum of the cross products (ignoring a factor of 2):

$$\sum_{k=0}^n u_k v_k = 0$$

By suitably regarding this sum while letting the dimension n become infinite, we are led to the corresponding integral

$$\int_0^1 u(k)v(k)\, dk$$

The kernel $K(\theta)$ of the integrand in the definition is a nonnegative weighting factor on the various components and adds no new complications. Notice, however, that the integral represents a noncountable number of dimensions.

A set of functions $g_n(\theta)$, $n = 0, 1, 2, \ldots$, is said to be orthogonal over the interval (a, b) with weight factor $K(\theta)$ if

$$\int_a^b K(\theta)g_m(\theta)g_n(\theta)\, d\theta = \begin{cases} 0, & \text{when } m \neq n \\ \lambda_n^2 & \text{when } m = n \end{cases} \qquad [K(\theta) \geq 0]$$

Of course, when $m = n$, the integrand is nonnegative and the integral must be a positive number. If $\lambda_n = 1$ for all n, then they are said to be *an orthonormal family* of functions. It is easy to convert an orthogonal set to an orthonormal set by merely dividing the nth function by the corresponding λ_n. Orthonormality is a convenient property in theoretical work because it eliminates the appearance of the λ_n. However, in practice, it is usually not done because it introduces awkward numerical factors.

Perhaps the best-known set of orthogonal functions is the Fourier set

$$1, \cos \theta, \cos 2\theta, \cos 3\theta, \ldots$$

$$\sin \theta, \sin 2\theta, \sin 3\theta, \ldots$$

over the interval $0 \leq \theta \leq 2\pi$ (or else $-\pi \leq \theta \leq \pi$). The weight factor $K(\theta) \equiv 1$.

To show that this set is orthogonal, we need to derive the following three integrals:

$$\int_0^{2\pi} \cos m\theta \cos n\theta \, d\theta = \begin{cases} 0, & m \neq n \\ \pi, & m = n \neq 0 \\ 2\pi, & m = n = 0 \end{cases}$$

$$\int_0^{2\pi} \cos m\theta \sin n\theta \, d\theta = 0$$

$$\int_0^{2\pi} \sin m\theta \sin n\theta \, d\theta = \begin{cases} 0, & m \neq n \\ \pi, & m = n \neq 0 \\ 0, & m = n = 0 \end{cases}$$

To derive the first, we use the trigonometric identity

$$\cos m\theta \cos n\theta = \tfrac{1}{2}[\cos(m + n)\theta + \cos(m - n)\theta]$$

Integrating, we get for $m \neq n$,

$$\frac{1}{2}\left[\frac{\sin(m + n)\theta}{(m + n)} + \frac{\sin(m - n)\theta}{(m - n)}\right]$$

When the limits 2π and 0 are inserted, this is clearly zero. When $m = n \neq 0$, we use the identity

$$\cos n\theta \cos n\theta = \cos^2 n\theta = \tfrac{1}{2}[1 + \cos 2n\theta]$$

Integrating, we get

$$\frac{1}{2}\left[\theta + \frac{\sin 2n\theta}{2n}\right]$$

which upon insertion of the limits gives the value π. Finally, when $m = n = 0$, the integrand clearly is the constant 1, and the integral therefore is 2π. The other orthogonality relations can be similarly verified by using the corresponding trigonometric identities.

Example. To normalize the trigonometric functions, we need to divide the sines and cosines by

$$\sqrt{\pi}$$

and the constant term 1 by

$$\sqrt{2\pi}$$

Exercises

4.2-1 Show that the cosines are orthogonal over the interval $0 \le \theta \le \pi$.

4.2-2 Show that the sines are orthogonal over the interval $0 \le \theta \le \pi$.

4.2-3 Complete the proof that the Fourier set of functions is orthogonal.

4.2-4 Show that the orthogonality applies also for any interval of length 2π, in particular, for $-\pi \le \theta \le \pi$.

4.2-5 Show that $g(\theta) = 0$ is orthogonal to all functions.

4.2-6 Show that the functions

$$1 \quad \cos 2\theta \quad \cos 4\theta \quad \cos 6\theta \ldots$$

are orthogonal over $0 \le \theta \le \pi/2$.

4.2-7 Show that $1, x, 2x^2 - 1$, are orthogonal over $-1 \le x \le 1$.

4.3 FORMAL EXPANSIONS

Given a function $g(\theta)$, $0 \le \theta \le 2\pi$, we assume that $g(\theta)$ has the formal expansion

$$g(\theta) = \frac{a_0}{2} + \sum_{k=1}^{\infty} [a_k \cos k\theta + b_k \sin k\theta] \tag{4.3-1}$$

The reason for the $a_0/2$ term will become clear shortly. To obtain an expression for the *coefficients* a_k, we multiply the equation on both sides by cos $m\theta$ and integrate $0 \leq \theta \leq 2\pi$. As a result of the orthogonality, we obtain

$$\int_0^{2\pi} g(\theta) \cos m\theta \, d\theta = \begin{cases} \pi a_m, & m \neq 0 \\ \pi a_0, & m = 0 \end{cases}$$

To get the b_k, we multiply by sin $m\theta$ in place of cos $m\theta$ and integrate

$$\int_0^{2\pi} g(\theta) \sin m\theta \, d\theta = \pi b_m \qquad (m \neq 0)$$

Thus the coefficients of the assumed expansion are given by the formulas

$$a_m = \frac{1}{\pi} \int_0^{2\pi} g(\theta) \cos m\theta \, d\theta \qquad (m = 0, 1, 2, \ldots)$$

$$b_m = \frac{1}{\pi} \int_0^{2\pi} g(\theta) \sin m\theta \, d\theta \qquad (m = 1, 2, 3, \ldots) \qquad (4.3\text{-}2)$$

These a_m and b_m are called the *Fourier coefficients* of the expansion of $g(\theta)$.

Notice that when the value of θ falls outside the original range $0 \leq \theta \leq 2\pi$, the function $g(\theta)$ is defined by the expansion to be periodic; thus we have $g(2\pi + \theta) = g(\theta)$ for all θ, and notice that any interval of length 2π will do, in particular $-\pi \leq \theta < \pi$.

Example 1. As an illustration of the expansion of a given function into a formal Fourier series, consider the rectangular pulse

$$g(\theta) = \begin{cases} \frac{1}{2}, & 0 \leq \theta < \pi \\ -\frac{1}{2}, & -\pi \leq \theta < 0 \end{cases}$$

Since

$$g(-\theta) = -g(\theta)$$

there will be no cosine terms in the final expansion. The coefficients of the sine terms are given by (4.3-2):

$$b_k = \frac{1}{\pi} \int_{-\pi}^{\pi} g(\theta) \sin k\theta \, d\theta$$

$$= \frac{1}{\pi} \int_0^{\pi} \sin k\theta \, d\theta$$

Doing the integration, we have

$$b_k = \frac{1}{\pi} \frac{[1 + (-1)^{k+1}]}{k} = \begin{cases} \dfrac{2}{\pi k}, & k \text{ odd} \\ 0, & k \text{ even} \end{cases}$$

Thus we have the formal expansion (4.3-1)

$$g(\theta) = \frac{2}{\pi} \left[\sin \theta + \frac{1}{3} \sin 3\theta + \frac{1}{5} \sin 5\theta + \cdots \right]$$

$$= \frac{2}{\pi} \sum_{k=0}^{\infty} \frac{1}{2k + 1} \sin(2k + 1)\theta$$

Figure 4.3-1 gives the graphs of the partial sums (labeled S_N) for 1, 5, and 9 terms of the series, for the interval $0 \le \theta < \pi$. For $-\pi < \theta \le 0$, the curves are the negatives of those shown. The figure indicates the quality of the approximation.

Example 2. For a second illustration, suppose that $g(\theta) = \theta$ and that we use the interval $-\pi \le \theta \le \pi$. Since the integrand is odd and the range of integration is symmetric about $\theta = 0$, we have (from 4.3-2)

$$a_m = \frac{1}{\pi} \int_{-\pi}^{\pi} \theta \cos m\theta \, d\theta = 0$$

For the b_m, we have

$$b_m = \frac{1}{\pi} \int_{-\pi}^{\pi} \theta \sin m\theta \, d\theta$$

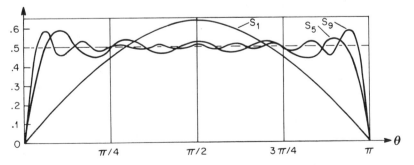

Figure 4.3-1 Partial Sums S_1, S_5, S_9 for Rectangular Pulse

Integrating by parts (recall that $\cos m\pi = (-1)^m$),

$$b_m = \frac{1}{\pi}\left[\frac{\theta(-\cos m\theta)}{m}\bigg|_{-\pi}^{\pi} + \frac{1}{m}\int_{-\pi}^{\pi}\cos m\theta\ d\theta \right]$$

$$= \frac{1}{\pi}\left[\frac{\pi}{m}2(-1)^{m+1} + \frac{1}{m}\frac{\sin m\theta}{m}\bigg|_{-\pi}^{\pi} \right]$$

$$= \frac{2}{m}(-1)^{m+1}$$

Thus we have the formal expansion (4.3-1)

$$\theta = 2\left[\sin\theta - \frac{\sin 2\theta}{2} + \frac{\sin 3\theta}{3} - \frac{\sin 4\theta}{4} + \cdots \right]$$

Figure 4.3-2 shows the first few partial sums (labeled S_N) as approximations to the function $g(\theta) = \theta$. Notice the effect of the periodicity at the ends of the interval, $-\pi \le \theta \le \pi$.

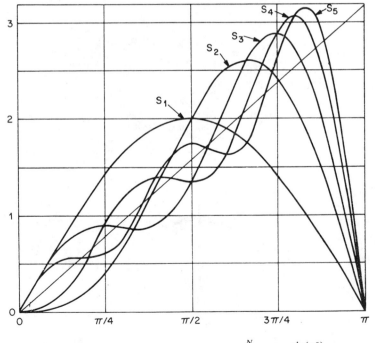

FIGURE 4.3-2 PARTIAL SUMS OF $S_N = -2\sum_{n=1}^{N}(-1)^n \dfrac{\sin(n\theta)}{n}$

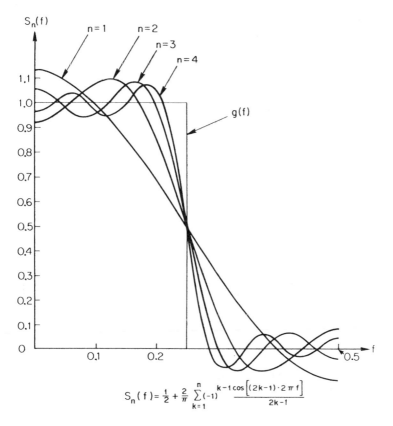

$$S_n(f) = \tfrac{1}{2} + \tfrac{2}{\pi} \sum_{k=1}^{n} (-1)^{k-1} \frac{\cos\left[(2k-1)\cdot 2\pi f\right]}{2k-1}$$

FIGURE 4.3-3

Example 3. As a third example consider the function (Figure 4.3-3) in the frequency notation of formula (2.1-1):

$$g(f) = \begin{cases} 0, & -\tfrac{1}{2} \le f < -\tfrac{1}{4} \\ 1, & -\tfrac{1}{4} < f < \tfrac{1}{4} \\ 0, & \tfrac{1}{4} < f \le \tfrac{1}{2} \end{cases}$$

Again, because $g(f) = g(-f)$, all the sine terms are zero. Next we need to convert the formula for the cosine coefficients (4.3-2):

$$a_k = \frac{1}{\pi} \int_{-\pi}^{\pi} g(\theta) \cos k\theta \ d\theta$$

to the f notation. Set $\theta = 2\pi f$. We get [note the change in meaning of $g(\theta)$]

$$a_k = 2 \int_{-1/2}^{1/2} \tilde{g}(f) \cos 2\pi k f \ df$$

For our particular function we get ($k \geq 1$), using symmetry,

$$a_k = 4 \int_0^{1/4} \cos 2\pi k f \, df$$

$$= 4 \left(\frac{\sin 2\pi k f}{2\pi k} \right) \Big|_0^{1/4} = \frac{2}{\pi k} \sin \frac{\pi}{2} k$$

For a_0 we have, similarly,

$$a_0 = 4 \int_0^{1/4} df = 1$$

Hence from (4.3-1)

$$\tilde{g}(f) = \frac{1}{2} + \frac{2}{\pi} \left[\cos 2\pi f - \frac{1}{3} \cos 6\pi f + \frac{1}{5} \cos 10\pi f - \cdots \right]$$

$$= \frac{1}{2} + \frac{2}{\pi} \sum_{n=0}^{\infty} (-1)^n \frac{\cos(2n + 1)2\pi f}{(2n + 1)}$$

The partial sums are shown in Figure 4.3-3.
We will return to these examples later.

Exercises

4.3-1 Write out the formulas for a Fourier expansion over the interval $0 \leq x \leq L$.

4.3-2 Prove that

$$g(\theta) = \begin{cases} 1 & 0 < \theta < \dfrac{L}{2} \\[2mm] -1 & \dfrac{L}{2} < \theta < L \end{cases}$$

is

$$g(\theta) = \frac{4}{\pi} \sum_{m=1}^{\infty} \frac{1}{2m - 1} \sin\left[\frac{(2m - 1)2\pi\theta}{L} \right]$$

4.3-3   Show that

$$g(\theta) = -\frac{1}{2} + \frac{\theta}{\pi} \quad (0 \le \theta \le \pi)$$

$$g(-\theta) = g(\theta)$$

is

$$g(\theta) = -4 \sum_{k=1}^{\infty} \frac{\cos(2k + 1)\theta}{\pi^2(2k + 1)^2}$$

hence (using $\theta = 0$)

$$\sum_{1}^{\infty} \frac{1}{k^2} = \frac{\pi^2}{6}$$

4.3-4   For the "rectified" function $(0 \le \theta < 2\pi)$ show that

$$g(\theta) = | \sin \theta | = \frac{2}{\pi} - \frac{4}{\pi} \sum_{1}^{\infty} \frac{1}{4m^2 - 1} \cos 2m\theta$$

4.3-5   The "half rectified" function

$$g(\theta) = \begin{cases} \sin \theta & 0 < \theta < \pi \\ 0 & -\pi < \theta < 0 \end{cases}$$

can be found from Example 3 and Exercise 4.3-4 using the observation

$$\text{half rect} = \frac{\text{orig} + \text{rect}}{2}$$

4.3-6   If $g(\theta) = \theta^2$, $(-\pi < \theta < \pi)$ show that

$$g(\theta) = \frac{\pi^2}{3} + 4 \sum_{k=1}^{\infty} (-1)^k \frac{\cos k\theta}{k^2}$$

4.3-7   For $g(\theta) = \theta(\pi - \theta)$, $(0 \le \theta < 2\pi)$ show that

$$g(\theta) = \frac{8}{\pi} \sum_{k=1}^{\infty} \frac{\sin(2k - 1)}{(2k - 1)^2}$$

4.3-8 For $g(\theta) = \sin n\theta$ (n not an integer) ($-\pi < \theta < \pi$) show that

$$g(\theta) = \frac{2 \sin \pi n}{\pi} \left(\frac{\sin \theta}{1 - n^2} - \frac{2 \sin 2\theta}{4 - n^2} + \frac{3 \sin 3\theta}{9 - n^2} - \cdots \right)$$

4.3-9 For $g(\theta) = e^{ax}$ ($-\pi < \theta < \pi$) find the Fourier expansion.

4.3-10 Expand $g(\theta) = e^{-a\theta}$ $|\theta| < \pi$ ($a > 0$).

4.4 ODD AND EVEN FUNCTIONS

Odd and even functions occur so often that their Fourier series are worth special examination. The fact that an arbitrary function $g(\theta)$ can be written as the sum of an odd function plus an even function simply by writing

$$g(\theta) = \frac{[g(\theta) - g(-\theta)]}{2} + \frac{[g(\theta) + g(-\theta)]}{2} \tag{4.4-1}$$

makes this point particularly true.

For an odd function, we see that $a_k = 0$ for all k, whereas

$$b_k = \frac{2}{\pi} \int_0^\pi g(\theta) \sin k\theta \, d\theta$$

Our differentiator filters (Section 3.5) were typical odd functions.

For an even function, we have

$$a_k = \frac{2}{\pi} \int_0^\pi g(\theta) \cos k\theta \, d\theta$$

and all the $b_k = 0$. The typical smoothing filter (Sections 3.2, 3.3, 3.6, and 3.7) is an even function.

By using additional suitable symmetries about $\theta = \pi/2$, we can obtain Fourier series with, say, only odd-indexed coefficients or only even-indexed coefficients. Other regular patterns of nonzero coefficients can also be obtained by suitably introducing proper odd and even symmetries into the definition of the function.

For example, consider the sawtooth function in Figure 4.4-1. It is an odd function, so we have only sine terms. It looks like a sine curve (slightly), but when we think about how it will integrate when multiplied by a sin 2θ, we see that the integral will be zero, and similarly for all even harmonics.

To check our opinion, we simply form the integral for the sine coefficients (note shift from θ to ω)

$$b_k = \frac{2}{\pi} \int_0^{\pi/2} \omega \sin k\omega \, d\omega + \frac{2}{\pi} \int_{\pi/2}^{\pi} (\pi - \omega) \sin k\omega \, d\omega$$

In the second integral, set $\pi - \omega = \omega'$:

$$b_k = \frac{2}{\pi} \int_0^{\pi/2} \omega \sin k\omega \, d\omega + \frac{2}{\pi} \int_0^{\pi/2} \omega' \sin k(\pi - \omega') \, d\omega'$$

Drop the primes and use

$$\sin k(\pi - \omega) = \sin k\pi \cos k\omega - \cos k\pi \sin k\omega$$
$$= (-1)^{k+1} \sin k\omega$$

Thus for k even the integrals will cancel, and for k odd, say $k = 2m + 1$,

$$b_{2m+1} = \frac{4}{\pi} \int_0^{\pi/2} \omega \sin(2m + 1)\omega \, d\omega$$

Integrate by parts:

$$b_{2m+1} = \frac{4}{\pi} \left\{ \omega \left(\frac{-\cos(2m + 1)\omega}{2m + 1} \right) \Bigg|_0^{\pi/2} + \int_0^{\pi/2} \frac{\cos(2m + 1)\omega}{2m + 1} \, d\omega \right\}$$

$$= \frac{4}{\pi} \left\{ \frac{\sin(2m + 1)\pi/2}{(2m + 1)^2} \right\} = (-1)^m \frac{4}{\pi(2m + 1)^2}$$

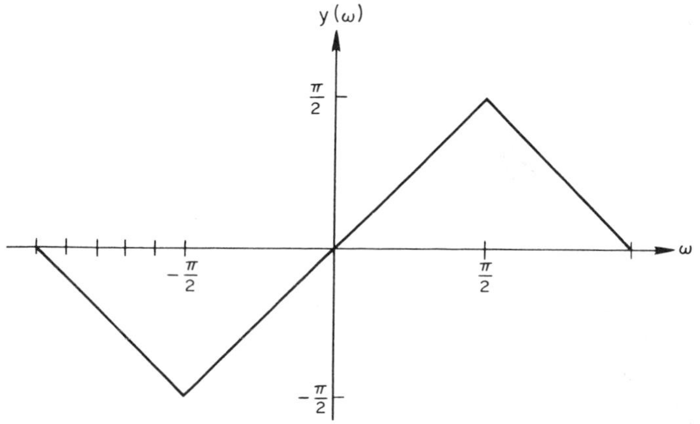

FIGURE 4.4-1 SAWTOOTH FUNCTION

Therefore

$$g(\omega) = \frac{4}{\pi} \left\{ \sin \omega - \frac{1}{9} \sin 3\omega + \frac{1}{25} \sin 5\omega - \cdots \right\} \qquad (4.4\text{-}2)$$

The first few terms are plotted in Figure 4.4-2 (using $f = \omega/2\pi$).

From this particular case it is easy to deduce how the general case will go. Define the arbitrary function in the interval $0 < \theta < \pi/2$. To get only cosine expansions you extend the function about the origin as an even function

$$g(\theta) = g(-\theta)$$

and for only sines as an odd function

$$g(\theta) = -g(-\theta)$$

To get odd harmonics for the cosines, you extend the function by picking odd symmetry about $\pi/2$,

$$g(\theta) = -g(\pi - \theta)$$

and for even harmonics you use

$$g(\theta) = g(\pi - \theta)$$

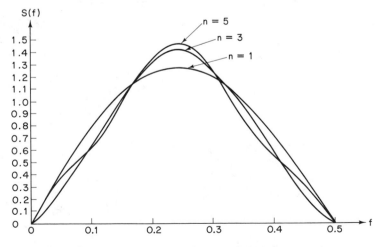

FIGURE 4.4-2 PARTIAL SUMS

To get the odd harmonics for the sine, you extend the function by picking even symmetry about $\pi/2$,

$$g(\theta) = g(\pi - \theta)$$

and for even harmonics you use

$$g(\theta) = -g(\pi - \theta)$$

It is simply a matter of looking at the terms you want to keep and using the appropriate symmetry. Of course, the idea can be indefinitely extended to get anything you want, but odd and even harmonics are as far as we will need to go.

Exercises

4.4-1   Expand $g(\theta) = e^{-a|\theta|}$, $|\theta| \leq L$.

4.4-2   Expand the function $g(\theta) = \theta^2$, $\theta \geq 0$, and $g(-\theta) = -g(\theta)$, $\theta \leq 0$, for the range $-L \leq \theta \leq L$.

4.4-3   Show that $g(\theta) = \theta$ has two expansions $(0 < \theta < \pi)$

$$\theta = \begin{cases} \pi - 2\left[\dfrac{\sin \theta}{1} + \dfrac{\sin 2\theta}{2} + \dfrac{\sin 3\theta}{3} + \cdots \right] \\[2em] \pi/2 - \dfrac{4}{\pi}\left[\dfrac{\cos \theta}{1^2} + \dfrac{\cos 3\theta}{3^2} + \dfrac{\cos 5\theta}{5^2} + \cdots \right] \end{cases}$$

4.4-4   For an even function in the interval $-2\pi \leq \theta \leq 2\pi$ and $g(2\pi - \theta) = g(\theta)$, show that $b_n = 0$ and

$$a_n = \begin{cases} \dfrac{2}{\pi} \displaystyle\int_0^\pi g(\theta) \cos \dfrac{n}{2} \theta \, d\theta, & n \text{ odd} \\[2em] 0, & n \text{ even} \end{cases}$$

4.4-5   Expand $g(\theta) = 1 - \theta^2/\pi^2$ $(|\theta| < \pi)$.

4.4-6   Expand $g(\theta) = \theta$ $(0 \leq \theta < \pi)$. Note that if you subtract π you get an odd function.

4.4-7   Show that for $|\theta| < \pi$

$$\theta \cos \theta = \frac{-\sin \theta}{2} + \frac{4 \sin 2\theta}{1.3} - \frac{6 \sin 2\theta}{3.5} + \frac{8 \sin 4\theta}{5.7} - \cdots$$

4.4-8   Expand $g(\theta) = \theta \sin \theta$ $(|\theta| < \pi)$.

4.5 FOURIER SERIES AND LEAST SQUARES

The Fourier series expansion is closely connected with the least-squares approximation of a function; indeed, we will show that the Fourier coefficients give the least-squares fit to the function. To do so, we set up the usual "sum of the squares of the residuals" for measuring the approximation of the function $g(\theta)$ by the nth partial sum $g_N(\theta)$, where

$$g_N(\theta) = \frac{A_0}{2} + \sum_{k=1}^{N} A_k \cos k\theta + \sum_{k=1}^{N} B_k \sin k\theta$$

Note the use of capital letters for the coefficients. We want to determine these coefficients so that

$$M = \int_{-\pi}^{\pi} \left[g(\theta) - \frac{A_0}{2} - \sum_{1}^{N} \{A_k \cos k\theta + B_k \sin k\theta\} \right]^2 d\theta$$

is a minimum. Therefore we differentiate with respect to the parameters A_m and B_m.

$$\frac{\partial M}{\partial A_m} = 0 \quad \Rightarrow \quad \int_{-\pi}^{\pi} g(\theta) \cos m\theta \, d\theta = \pi A_k$$

$$\frac{\partial M}{\partial B_m} = 0 \quad \Rightarrow \quad \int_{-\pi}^{\pi} g(\theta) \sin m\theta \, d\theta = \pi B_k$$

For finite N we have therefore proved that the formal Fourier expansion is the least-squares fit.

If the equality

$$\frac{1}{\pi} \int_{-\pi}^{\pi} g^2(\theta) \, d\theta = \frac{a_0^2}{2} + \sum_{k=1}^{\infty} [a_k^2 + b_k^2] \qquad (4.5\text{-}1)$$

occurs, it is called *Parseval's equality*.

The inequality which occurs for finite N is called *Bessel's inequality*:

$$\frac{1}{\pi} \int_{-\pi}^{\pi} g^2(\theta) \, d\theta \geq \left\{ \frac{a_0^2}{2} + \sum_{k=1}^{N} (a_k^2 + b_k^2) \right\} \qquad (4.5\text{-}2)$$

Bessel's inequality is very useful in estimating the mean square error of the fit as we gradually increase the number of coefficients being computed. When rearranged, this equation says that the sum of the squares of the residuals of the fit (approximation) is the difference between the integral of the square of the function (with a factor of $1/\pi$) and the sum of the squares of the coefficients that have been computed. (Remember to take $a_0^2/2$.)

It is clear from Bessel's inequality that the series

$$\frac{a_0^2}{2} + \sum_{k=1}^{\infty} \{a_k^2 + b_k^2\}$$

is bounded above and therefore converges, provided that the functions $g(\theta)$ and $g^2(\theta)$ are integrable. Thus the Fourier coefficients of the expansion

$$a_k = \frac{1}{\pi} \int_{-\pi}^{\pi} g(\theta) \cos k\theta \ d\theta \to 0$$

$$b_k = \frac{1}{\pi} \int_{-\pi}^{\pi} g(\theta) \sin k\theta \ d\theta \to 0 \qquad (4.5\text{-}3)$$

as k approaches infinity. We will later need a corollary to this result

$$\frac{1}{\pi} \int_{-\pi}^{\pi} g(\theta) \sin\left(k + \frac{1}{2}\right)\theta \ d\theta \to 0 \qquad (4.5\text{-}4)$$

as k approaches infinity.

To prove this corollary we expand $\sin(k + \frac{1}{2})\theta = \sin k\theta \cos \theta/2 + \cos k\theta \sin \theta/2$. Now if $g(\theta)$ meets the necessary conditions for a Fourier expansion then so do

$$g(\theta) \cos \frac{\theta}{2} \equiv g_1(\theta)$$

$$g(\theta) \sin \frac{\theta}{2} \equiv g_2(\theta)$$

Hence we have

$$\int_{-\pi}^{\pi} g_1(\theta) \sin k\theta \ d\theta + \int_{-\pi}^{\pi} g_2(\theta) \cos k\theta \ d\theta$$

$$= b_k[\text{of } g_1(\theta)] + a_k[\text{of } g_2(\theta)] \to 0$$

because both separately $\to 0$ by (4.5-3).

Exercises

4.5-1 Apply Bessel's inequality to the first few terms of the first example in Section 4.3.

4.5-2 Do the same for the second example.

4.5-3 Apply Bessel's inequality to Exercise 4.3-4.

4.5-4 Do the same for Exercise 4.3-5.

4.5-5 Do the same for Exercise 4.3-6.

4.6 CLASS OF FUNCTIONS AND RATE OF CONVERGENCE

Under the assumption that $g(\theta)$ and its square are both integrable, we have found that the Fourier coefficients a_k and b_k both approach zero as k approaches infinity. This result does not prove that the series converges, much less that, if it does converge, it will approach the original function $g(\theta)$. So let us investigate the rate of convergence and then examine what the series approaches.

For our applications, which class of functions should be considered? In almost all applied problems we need at worst a function consisting of a finite number of pieces (adjacent intervals) such that in each interval all the required derivatives exist. Thus we use piecewise analytic functions. This piecewise property means that, if we wish to be very careful, we must speak of left-hand and right-hand derivatives at the ends of the intervals. However, we shall assume the reader's awareness. This class of functions allows jumps, corners, and similar features: jumps in the function if the function is discontinuous, corners in the function if the first derivative is discontinuous, sudden changes in the curvature if the second derivative is discontinuous, and so on.

Given a function of this class, what are its Fourier coefficients? We have for J pieces (intervals)

$$a_k = \frac{1}{\pi} \int_{-\pi}^{\pi} g(\theta) \cos k\theta \, d\theta$$

$$= \frac{1}{\pi} \sum_{j=1}^{J} \int_{\theta_{j-1}}^{\theta_j} g(\theta) \cos k\theta \, d\theta$$

where $\theta_0 = -\pi$, $\theta_J = \pi$, and the other θ_j's mark the ends of the pieces where the breaks in the function occur. A similar formula applies for the b_k.

Next, integrating by parts, we have

$$a_k = \frac{1}{\pi} \sum_{j=1}^{J} \left[g(\theta) \left(\frac{\sin k\theta}{k} \right) \Big|_{\theta_{j-1}}^{\theta_j} - \int_{\theta_{j-1}}^{\theta_j} g'(\theta) \left(\frac{\sin k\theta \, d\theta}{k} \right) \right]$$

If the function is continuous [remember that this includes the condition that $g(-\pi) = g(\pi)$] then because the value from the top of one interval is exactly the same as the value from the bottom of the next interval all the integrated terms will cancel out independent of the value of k. Thus we will be left with only the new integral. However, if the function is not continuous the terms will not cancel out for all k and the a_k will in general be of order $1/k$.

If the cancellation does occur then we can integrate by parts again to get

$$a_k = \frac{1}{\pi} \sum_{j=1}^{J} {}' \left[g'(\theta) \left(\frac{\cos k\theta}{k^2} \right) \Big|_{\theta_{j-1}}^{\theta_j} - \int_{\theta_{j-1}}^{\theta_j} \frac{g''(\theta) \cos k\theta}{k^2} d\theta \right]$$

This time it is the continuity of the first derivative that is necessary if the integrated part is to cancel out. Continuing in this way, we find that without continuity of the function the coefficients of the Fourier expansion have some terms like $1/k$, that without continuity of the first derivative the coefficients are like $1/k^2$, that without continuity of the second derivative they are like $1/k^3$, and so forth. Similar statements are true for the b_k coefficients.

Once the coefficients are of order $1/k^2$, then, since the trigonometric functions are not greater than 1 in size, we have the convergence of the Fourier series. Only in the case of a discontinuous function need we look more closely.

Exercises

What is the rate of decrease of the Fourier coefficients if the function is for $|\theta| < \pi$:

4.6-1 $g(\theta) = \begin{cases} 1 & 0 \le \theta < \pi \\ -1 & -\pi \le \theta < 0 \end{cases}$

4.6-2 $g(\theta) = |\theta|$

4.6-3 $g(\theta) = 1 - \left(\dfrac{\theta}{\pi} \right)^2$

4.6-4 $g(\theta) = 3 \left(\dfrac{\theta}{\pi} \right)^2 - 2 \left(\dfrac{\theta}{\pi} \right)^3$ $(0 \le \theta < \pi)$

4.6-5 $g(\theta) = e^\theta$

4.6-6 $g(\theta) = e^{-|\theta|}$

4.6-7 $g(\theta) = \sin \theta$

4.7 CONVERGENCE AT A POINT OF CONTINUITY

In this section we consider the convergence of the formal Fourier series. Specifically, we ask: Does the series converge at a point *inside* one of the intervals $\theta_j \le \theta \le \theta_{j+1}$, *and*, if so does it approach the function at the point of continuity? In the next section we will discuss the convergence at points of discontinuity.

We have, for the sum of the frequencies up to N, the partial sum

$$g_N(\theta) = \frac{a_0}{2} + \sum_{k=1}^{N} [a_k \cos k\theta + b_k \sin k\theta] \qquad (4.7\text{-}1)$$

where, of course, the Fourier coefficients are given by

$$a_k = \frac{1}{\pi} \int_{-\pi}^{\pi} g(s) \cos ks \, ds$$

$$b_k = \frac{1}{\pi} \int_{-\pi}^{\pi} g(s) \sin ks \, ds$$

The dummy variable of integration s has been used in order to avoid confusion. Eliminating the coefficient a_k and b_k in the partial sum (4.7-1), we have, on interchanging the finite summation and integration operations,

$$g_N(\theta) = \frac{1}{\pi} \int_{-\pi}^{\pi} g(s) \left[\frac{1}{2} + \sum_{k=1}^{N} \{\cos ks \cos k\theta + \sin ks \sin k\theta\} \right] ds$$

or

$$g_N(\theta) = \frac{1}{\pi} \int_{-\pi}^{\pi} g(s) \left[\frac{1}{2} + \sum_{k=1}^{N} \cos k(s - \theta) \right] ds$$

For convenience, we assume that $g(\theta)$ is periodic; therefore, by setting

$$s - \theta = u \qquad (4.7\text{-}2)$$

we shift the coordinate system and place the point θ (where we are examining the convergence question) in the middle of the interval of integration in the variable u. The result is

$$g_N(\theta) = \frac{1}{\pi} \int_{-\pi}^{\pi} g(\theta + u) \left[\frac{1}{2} + \sum_{k=1}^{N} \cos ku \right] du \qquad (4.7\text{-}3)$$

To sum the expression in the brackets,

$$s_N = \frac{1}{2} + \sum_{k=1}^{N} \cos ku \tag{4.7-4}$$

we multiply through by $\sin u/2$ and apply some elementary trigonometry to get

$$\left(\sin \frac{u}{2}\right)s_N = \frac{1}{2}\left[\sin \frac{u}{2} + \left(\sin \frac{3u}{2} - \sin \frac{u}{2}\right) + \left(\sin \frac{5u}{2} - \sin \frac{3u}{2}\right) + \cdots \right.$$
$$\left. + \left(\sin \frac{(2N+1)u}{2} - \sin \frac{(2N-1)u}{2}\right)\right]$$
$$= \frac{1}{2} \sin \frac{(2N+1)u}{2}$$

Because of the cancellation of terms, only the last one remains; and we have for the sum of the cosines

$$s_N = \frac{\sin(2N+1)u/2}{2 \sin u/2} \tag{4.7-5}$$

Therefore, the partial sum of the Fourier series is

$$g_N(\theta) = \frac{1}{\pi} \int_{-\pi}^{\pi} g(\theta + u) \frac{\sin\{(2N+1)u/2\}}{\sin u/2} \frac{du}{2} \tag{4.7-6}$$

To prove the convergence of the partial sums to the function $g(\theta)$ and that it approaches the function $g(\theta)$, we need an expression for the difference

$$g_N(\theta) - g(\theta)$$

To get it, we note that integrating both expressions (4.7-4) and (4.7-5) for s_N from $-\pi$ to π and dividing by π gives

$$1 = \frac{1}{\pi} \int_{-\pi}^{\pi} \left[\frac{\sin(2N+1)u/2}{\sin u/2} \right] \frac{du}{2}$$

Multiplying by $g(\theta)$ and using the fact that the integration is with respect to u, we get the result

$$g(\theta) = \frac{1}{\pi} \int_{-\pi}^{\pi} [g(\theta)] \left[\frac{\sin(2N+1)u/2}{\sin u/2} \right] \frac{du}{2} \tag{4.7-7}$$

Subtracting this result (4.7-7) from the formula for the partial sum (4.7-6) provides the needed expression

$$g_N(\theta) - g(\theta) = \frac{1}{\pi} \int_{-\pi}^{\pi} [g(\theta + u) - g(\theta)] \left[\frac{\sin(2N + 1)u/2}{\sin u/2} \right] \frac{du}{2}$$

To prove the convergence of the partial sum $g_N(\theta)$ to $g(\theta)$ at the point θ, it is necessary to show that this difference approaches zero as N increases. To do so, we simply insert a factor u in both the numerator and the denominator and then rearrange the expression:

$$g_N(\theta) - g(\theta) = \frac{1}{\pi} \int_{-\pi}^{\pi} \left[\frac{\{g(\theta + u) - g(\theta)\}}{u} \right] \left[\frac{\sin(2N + 1)u/2}{(\sin u/2)/(u/2)} \right] du$$

In this form the first bracket approaches the derivative of $g(\theta)$ as u approaches 0, so this term gives no trouble. The denominator of the second bracket has the limit 1 as u approaches 0 and is well behaved elsewhere in the range of integration. Thus the function

$$\phi(u) = \frac{g(\theta + u) - g(\theta)}{2 \sin u/2}$$

is well behaved; and if it is integrable square, we can apply the earlier corollary (4.5-4) to the integral

$$\frac{1}{\pi} \int_{-\pi}^{\pi} \phi(u) \sin \left[\frac{(2N + 1)u}{2} \right] du$$

Clearly, as N approaches infinity, the difference between the partial sum and the function approaches zero at any point at which a two-sided derivative exists. Thus we have convergence to the function.

THEOREM.   *The formal Fourier series converges to the function at any point where the derivative exists [provided that $\phi^2(u)$ is integrable].*

Example.   Consider the function

$$g(\theta) = |\theta| \qquad (|\theta| < \pi)$$

This is the "typical corner" function. We have

$$a_k = \frac{1}{\pi} \int_{-\pi}^{\pi} |\theta| \cos k\theta \, d\theta$$

$$= \frac{2}{\pi} \int_0^\pi \theta \cos k\theta \, d\theta$$

$$= \frac{2}{\pi} \left\{ \theta \frac{\sin k\theta}{k} \Big|_0^\pi - \int_0^\pi \frac{\sin k\theta}{k} \, d\theta \right\}$$

$$= \frac{2}{\pi} \frac{\cos k\theta}{k^2} \Big|_0^\pi = \frac{2}{\pi k^2} [(-1)^k - 1]$$

$$= \begin{cases} \dfrac{-4}{\pi k^2} & \text{if} \quad k \text{ is odd} \\ 0 & \text{if} \quad k \text{ is even} \end{cases}$$

$$a_\theta = \frac{2}{\pi} \int_0^\pi \theta \, d\theta = \frac{2}{\pi} \frac{\pi^2}{2} = \pi$$

Therefore

$$g(\theta) = |\theta| = \frac{\pi}{2} - \frac{4}{\pi} \sum_0^\infty \frac{\cos(2m + 1)\theta}{(2m + 1)^2}$$

At $\theta = 0$ we have for the partial sums

$$g_N(0) = \frac{\pi}{2} - \frac{4}{\pi} \sum_0^N \frac{1}{(2m + 1)^2}$$

At the corner the remainder

$$R_N = \sum_{N+1}^\infty \frac{1}{(2m + 1)^2} \sim \int_{N+1}^\infty \frac{dx}{(2x + 1)^2} = \frac{1}{2N + 3}$$

and the approach is slow.

Exercise

4.7-1   Analyze in detail the rate of fall off of the coefficients and the rate of convergence at $\theta = 0$ for

$$g(\theta) = \begin{cases} \theta & 0 \le \theta < \pi \\ 0 & -\pi < \theta \le 0 \end{cases}$$

Note: Example 2 of Section 4.3 plus the example of this section will give you the coefficients.

4.8 CONVERGENCE AT A POINT OF DISCONTINUITY

At a point of discontinuity θ_j, that is, at the end of some interval being used, we do not have a suitable derivative—only one-sided derivatives— and the preceding argument breaks down.

The first obvious step is to break the integral into two halves, since the function is different on the two sides of a point θ_j. As a result of the shift in the coordinate system (4.7-2), the point θ_j is now at 0. We can write

$$g_N(\theta) = \frac{1}{\pi} \int_{-\pi}^{0} + \frac{1}{\pi} \int_{0}^{\pi}$$

In the first integral we replace u by $-u$ and then combine both integrals corresponding to (4.7-6):

$$g_N(\theta) = \frac{1}{\pi} \int_{0}^{\pi} [g(\theta_j + u) + g(\theta_j - u)] \frac{\sin(2N + 1)u/2}{\sin u/2} \frac{du}{2}$$

In the earlier proof, when at a point of continuity, we subtracted out $g(\theta)$ using (4.7-7). What must we do now? Clearly, we need something for both halves, which would be

$$\frac{g(\theta_j +) + g(\theta_j -)}{2}$$

(the + and − signs mean limit from the right and left, respectively). This is the average of the two limiting values (the factor one-half occurs because the range of integration is now one-half of what it was previously) and it fits with the one-sided derivatives. Using this value, we obtain an expression for the difference:

$$g_N(\theta_j) - \left[\frac{g(\theta_j +) + g(\theta_j -)}{2} \right]$$

which will approach zero as before (although we need an extra argument to prove that the corollary can be suitably modified for the half-interval). Thus we have the following theorem.

THEOREM. *At a point of discontinuity the formal Fourier series converges to the average of the two limiting values of the function [provided that $\phi^2(u)$ is integrable].*

We have seen this effect in the three examples of Section 4.3.

It is probably worth pointing out that the *convergence* of a Fourier series at a point is a local property, whereas the *rate of convergence* is a global property.

Exercise

4.8-1 Fill in the details of this section.

4.9 THE COMPLEX FOURIER SERIES

In Section 2.4 we observed that the complex form of the Fourier series is much easier to handle and to understand (in the long run) than is the real form using sines and cosines. It is time that we shifted to the complex notation.

By adding and subtracting Euler's identities $[i - \sqrt{(-1)}]$, (2.4-4)

$$e^{i\theta} = \cos \theta + i \sin \theta$$

$$e^{-i\theta} = \cos \theta - i \sin \theta$$

we get (2.4-6)

$$\cos \theta - \frac{e^{i\theta} + e^{-i\theta}}{2}$$

$$\sin \theta = \frac{e^{i\theta} - e^{-i\theta}}{2i}$$

Since $\sin \theta$ and $\cos \theta$ are linearly independent, so are $e^{i\theta}$ and $e^{-i\theta}$. Thus, corresponding to the two functions $\sin \theta$ and $\cos \theta$, we have two complex functions $e^{i\theta}$ and $e^{-i\theta}$; a single real frequency will give rise to two frequencies in the complex notation, one positive and one negative.

In this complex notation the Fourier series for the interval $-\pi \leq \theta \leq \pi$,

$$g_N(\theta) = \frac{a_0}{2} + \sum_{k=1}^{N} [a_k \cos k\theta + b_k \sin k\theta]$$

becomes

$$g_N(\theta) = \sum_{k=-N}^{k=N} c_k e^{ik\theta}$$

where

$$c_k = \frac{a_k - ib_k}{2} \qquad \text{for } k > 0$$

$$c_0 = \frac{a_0}{2}$$

$$c_k = \frac{a_k + ib_k}{2} \qquad \text{for } k < 0 \qquad (4.9\text{-}1)$$

If $g(\theta)$ is an even function in θ, then, by Section 4.4, $b_k = 0$ and $c_k = a_k/2$. If $g(\theta)$ is an odd function, $a_k = 0$ and $c_k = \pm ib_k/2$.

Instead of this indirect proof, the complex form of the Fourier series can be derived directly by noting that the integral

$$\int_{-\pi}^{\pi} e^{ik\theta} e^{-im\theta} \, d\theta = \begin{cases} 0, & k \neq m \\ 2\pi, & k = m \end{cases} \qquad (4.9\text{-}2)$$

So if we assume the formal expansion

$$g_N(\theta) = \sum_{k=-N}^{k=N} c_k e^{ik\theta} \qquad (4.9\text{-}3)$$

the coefficient c_m can be found by multiplying both sides by $e^{-im\theta}$, which is the complex conjugate of $e^{im\theta}$, and integrating:

$$\int_{-\pi}^{\pi} e^{-im\theta} g_N(\theta) \, d\theta = 2\pi c_m \qquad (4.9\text{-}4)$$

It is easy to see (by replacing i by $-i$ if in no other way) that for a real function $g_N(\theta)$ we will have

$$c_k = \overline{c_{-k}}$$

where the overbar means the complex conjugate. Furthermore, the complex Fourier series is merely a notational change, a change, however, that greatly simplifies the notation. The mental adjustment necessary (thinking about complex functions with both positive and negative frequencies) is well worth the gain in the simple algebra that results.

In Section 4.5 we derived Bessel's inequality (4.5-2) for real Fourier series expansions. For the complex expansion, we proceed in the same fashion *except* that we use a factor of $1/2\pi$ in the averaging process.

$$\frac{1}{2\pi} \int_{-\pi}^{\pi} \left| g(\theta) - \sum_{k=-N}^{k=N} c_k e^{ik\theta} \right|^2 d\theta \geq 0$$

We expand the absolute value squared as before, remembering that it is the product of the function times its conjugate. We then replace what integrals we can by the corresponding complex Fourier coefficients c_k and rearrange in order to obtain finally

$$\frac{1}{2\pi} \int_{-\pi}^{\pi} |g(\theta)|^2 \, d\theta \geq \sum_{k=-N}^{k=N} |c_k|^2 \qquad (4.9\text{-}5)$$

From Parseval's equality (4.5-1), if it holds, we see that the sum of the squares of the coefficients (which measures the noise propagation through a nonrecursive filter; Section 1.7) can be found from the integral of the square of the absolute value of the function.

If we return to the fundamental sampling effects discussed in Chapter 2, we still require that the complex Fourier series have two samples in the highest frequency present (if we are to avoid aliasing of some high frequencies into lower frequencies when we do the sampling). The Nyquist frequency interval still goes from $-\pi$ to π.

We have used radian measure (except in plotting) because of its convenience in the calculus operations. As noted, for most design purposes it is more convenient to measure angles in rotations. Thus we often make the change of notation (2.1-1)

$$\theta = \omega = 2\pi f$$

The Nyquist folding frequency in rotations is at $\frac{1}{2}$ rotation, and the basic interval of frequencies runs from $-\frac{1}{2}$ to $\frac{1}{2}$. If we are concerned with sampling rates, then the Nyquist interval is from $-\frac{1}{2}$ Hz (hertz) to $\frac{1}{2}$ Hz, *measured in time units for which the sample spacing is 1*. Note that in all the various notations at least two samples are always needed in the highest frequency present.

To see how the Fourier expansion is relevant to the transfer function, we convert the Fourier series from the independent variable θ to the independent variable ω (or f). Given a function of ω, say $g(\omega)$, the corresponding expansion is

$$g(\omega) = \sum_{k=-\infty}^{\infty} c_k e^{ik\omega} \qquad (4.9\text{-}6)$$

where the c_k are given by

$$c_k = \frac{1}{2\pi} \int_{-\pi}^{\pi} g(\omega) e^{-ik\omega} \, d\omega \qquad (4.9\text{-}7)$$

This notation matches that for the eigenvalue $\lambda(\omega)$ of Section 2.5,

$$\lambda(\omega) = \sum_{k=-N}^{N} c_k e^{-i\omega k}$$

For if we set $k = -k'$, we have

$$\lambda(\omega) = \sum_{k'=-N}^{N} c_{-k'} e^{i\omega k'}$$

Beginning with Section 3.2, we used the transfer function notation

$$\lambda(\omega) = H(\omega) = \sum_{k=-N}^{N} c_{-k} e^{i\omega k}$$

The fact that the subscript on the coefficients was negative was generally concealed by the symmetry of the formulas or by avoiding labeling the coefficients of a particular formula with their abstract symbols c_{-k}. *Thus for a symmetric formula the c_k that occur in the complex Fourier expansion of the transfer function are the same as the c_k that occur in the original definition of the digital filter. For an antisymmetric formula they have the opposite signs in the subscripts.*

For convenience, we write out the formal Fourier series in the f notation:

$$\tilde{g}_N(f) = \sum_{k=-N}^{N} c_k e^{2\pi i k f}$$

where

$$c_k = \int_{-1/2}^{1/2} \tilde{g}_N(f) e^{-2\pi i k f} \, df \tag{4.9-8}$$

In the "real" notation we have, for the variable f,

$$\tilde{g}_N(f) = \frac{a_0}{2} + \sum_{k=1}^{k=N} [a_k \cos 2\pi k f + b_k \sin 2\pi k f] \tag{4.9-9}$$

where

$$a_k = 2 \int_{-1/2}^{1/2} \tilde{g}(f) \cos 2\pi k f \, df$$

$$b_k = 2 \int_{-1/2}^{1/2} \tilde{g}(f) \sin 2\pi k f \, df \tag{4.9-10}$$

Exercises

4.9-1 Find the Fourier expansion of $g(\theta) = \sin^5 \theta$. (You do not need to do any integrations.) ($|\theta| < \pi$). *Answer:* $\sin^5(\theta) = 5 \sin \theta - 20 \sin 3\theta + 16 \sin 5\theta$

4.9-2 Do the same for $g(\theta) = \sin^4 \theta$.

4.9-3 Using complex notation expand

$$g(\theta) = e^{a\theta} \qquad (|\theta| < \pi)$$

4.9-4 Using complex notation expand

$$g(\theta) = e^{-|\theta|} \qquad (|\theta| \le \pi)$$

Convert to real functions.

4.9-5 Expand $g(f) = e^{2\pi i f}$. (You have the answer!)

4.9-6 Expand $g(f) = e^{-|2\pi f|}$. ($|f| \le \frac{1}{2}$)

4.9-7 Expand $g(\theta) = |\sin \theta|$.

4.10 THE PHASE FORM OF A FOURIER SERIES

In Section 2.4 we found that although the individual coefficients a_k and b_k of a Fourier series depend for their values on the origin taken for the periodic function $f(\theta)$, the quantity

$$a_k^2 + b_k^2$$

is invariant under coordinate translations. For this reason, the *phase form* of a Fourier series is useful. To obtain it, we start with the usual series

$$g(\theta) = \frac{a_0}{2} + \sum_{k=1}^{\infty} [a_k \cos k\theta + b_k \sin k\theta]$$

and write it in the form

$$g(\theta) = \frac{a_0}{2} + \sum_{k=1}^{\infty} \sqrt{a_k^2 + b_k^2} \left[\frac{a_k}{\sqrt{a_k^2 + b_k^2}} \cos k\theta + \frac{b_k}{\sqrt{a_k^2 + b_k^2}} \sin k\theta \right]$$

If we define A_k and ϕ_k such that

$$\sqrt{a_k^2 + b_k^2} = A_k \tag{4.10-1}$$

$$\frac{a_k}{A_k} = \cos \phi_k$$

$$\frac{b_k}{A_k} = -\sin \phi_k \tag{4.10-2}$$

we have

$$g(\theta) = \frac{a_0}{2} + \sum_{k=1}^{\infty} A_k \cos(k\theta + \phi_k)$$

which exhibits the phase ϕ_k and the amplitude A_k in a clear form. Each pair a_k and b_k is equivalent to the pair A_k and ϕ_k.

In the complex notation

$$g(\theta) = \sum_{k=-\infty}^{\infty} c_k e^{ik\theta}$$

we merely write the coefficients c_k in the polar form

$$c_k = A_k e^{i\phi_k}$$

and we have the corresponding phase form

$$g(\theta) = \sum_{k=-\infty}^{\infty} A_k e^{i(k\theta + \phi_k)}$$

A closely related form of the Fourier series is the *delay form*,

$$f(\theta) = \sum_{k=-\infty}^{\infty} A_k e^{ik(\theta + \tau_k)}$$

where we have written

$$k\tau_k = \phi_k$$

In this notation the shift θ_0 of the time origin merely adds a fixed amount θ_0 to each delay term τ_k. In the phase notation the shift produces a $\theta_0 k$

addition to the phase ϕ_k. A constant delay τ_0 maintains the shape of a time signal.

When determining ϕ or ϕ_k use 4.10-2 and *not*

$$\tan \phi = \frac{\sin \phi}{\cos \phi} = \frac{-b_k}{a_k}$$

since you may lose a factor of π.

Exercises

4.10-1   Put the half rectified sinusoid (Exercise 4.3-5) into the phase form.

4.10-2   Put the expansion of

$$g(\theta) = e^{\theta}$$

into phase form.

4.10-3   Put

$$g(\theta) = \sin \theta + \cos \theta + \tfrac{1}{4}(\sin 2\theta - \cos 2\theta)$$
$$+ \tfrac{1}{9}(\sin 3\theta + \cos 3\theta) + \tfrac{1}{16}(\sin 4\theta - \cos 4\theta) - \cdots$$

into phase form.

5

Windows

5.1 INTRODUCTION

The concept of a *window* is probably the single most confusing one in the whole field of filter design. The difficulty is that windows come in pairs; for the Fourier series there is both the function and the corresponding set of coefficients. Given either form, there is a unique corresponding window of the other form. Chapter 3 had many examples of: given the coefficients, what function has them? The answer was, of course, the transfer function $H(\omega)$. In Chapter 4 we did the reverse; we started with a function and found its Fourier coefficients.

But further confusion arises from the fact that both forms can be used either as a multiplying or as a convolving window, and in either the frequency or time domains. It is for this reason that we have used the neutral variable θ in developing the theory of the Fourier series.

In Section 1.1 we saw that a nonrecursive digital filter (1.1-1) is the convolution of a set of weighting coefficients c_k with a set of data u_{n-k}. We see the data through the window of coefficients, and this window is *translucent* in the sense that we get to see only the sum of the products, but not any single product. On the other hand, we will have multiplying windows when we truncate a Fourier series; we are in effect multiplying the infinite array of Fourier coefficients by a block of 1s and all the others (that we discard) by 0s (Figure 3.4-1). In this kind of a window each product is seen separately; the window is *transparent*. Similarly, if we record a finite run of data of a continuous function we are windowing the input sequence of numbers by a multiplying window of 1 where we record and 0 where we do not record.

The basic connections between the two kinds of windows are given in the next section on the two convolution theorems. Later there will be further convolving theorems for other situations than the Fourier series.

A very important effect in the theory of Fourier series is the Gibbs phenomenon. Chapter 4 showed that for reasonable functions the formal Fourier series converges to the function at a point of continuity and converges to the average of the two limits at a point of discontinuity. Each term of the Fourier series is continuous; consequently, the theorem "a *uniformly* convergent series of continuous functions converges to a continuous function" implies that at a point of discontinuity the convergence of a Fourier series cannot be uniform; something peculiar must happen. What occurs is a significant overshoot no matter how many terms of the series are taken; it is called the Gibbs phenomenon. For those unfamiliar with the idea of uniform convergence, the subject is discussed briefly in Appendix 5.A of this chapter.

5.2 GENERATING NEW FOURIER SERIES:
THE CONVOLUTION THEOREMS

The basic method for finding the Fourier series of a given function $g(\theta)$, $-\pi < \theta < \pi$,

$$g(\theta) = \sum_{k=-\infty}^{k=\infty} c_k e^{ik\theta}$$

is to compute the coefficients from the integrals

$$c_k = \frac{1}{2\pi} \int_{-\pi}^{\pi} g(\theta) e^{-ik\theta} \, d\theta$$

This process can be both time- and effort-consuming for many functions.

How can we, from known (comparatively simple) expansions, derive other expansions? Evidently, if we know both

$$g(\theta) = \sum_{k=-\infty}^{k=\infty} c_k e^{ik\theta}$$

$$h(\theta) = \sum_{k=-\infty}^{k=\infty} d_k e^{ik\theta}$$

then we know

$$Ag(\theta) + Bh(\theta) = \sum_{k=-\infty}^{k=\infty} [Ac_k + Bd_k]e^{ik\theta}$$

How about the expansion of the product of two functions? For *real-valued* functions, we have

$$g(\theta)h(\theta) = \sum_{k=-\infty}^{k=\infty} c_k e^{ik\theta} \sum_{m=-\infty}^{m=\infty} d_m e^{im\theta}$$

Setting $n = k + m$ and rearranging terms, we have

$$g(\theta)h(\theta) = \sum_{n=-\infty}^{n=\infty} e^{in\theta} \left[\sum_{k=-\infty}^{k=\infty} c_k d_{n-k} \right]$$

The nth coefficient in the expansion of the product of the two functions is, therefore,

$$\sum_{k=-\infty}^{k=\infty} c_k d_{n-k} = \sum_{k=-\infty}^{k=\infty} d_k c_{n-k}$$

This is called *the convolution* of the sequence c_k by the sequence d_k. Thus, to get the coefficients for the expansion of a product of two given functions, we compute the convolutions of the coefficients.

In the z-transform notation the Fourier series are

$$g(\theta) = \sum_{n=-\infty}^{\infty} c_k z^k$$

$$h(\theta) = \sum_{n=-\infty}^{\infty} d_k z^k$$

$$g(\theta)h(\theta) = \sum_{k=-\infty}^{\infty} c_k z^k \sum_{n=-\infty}^{\infty} d_m z^m$$

$$= \sum_{n=-\infty}^{\infty} \left[\sum_{k=-\infty}^{\infty} c_k d_{n-k} \right] z^n$$

This form resembles the multiplication of two power series—except that we have used both positive and negative powers.

For *complex-valued* functions, it is customary to use the product of one function times the complex conjugate of the other.

The convolution that results from multiplying two functions together suggests asking which function corresponds to multiplying the corresponding coefficients of the two expansions; that is, to which function does the series

$$\sum c_k d_k e^{ik\theta}$$

correspond? To answer, we introduce *the convolution of two periodic functions g(θ) and h(θ)* as

$$m(\theta) = \frac{1}{2\pi} \int_{-\pi}^{\pi} g(s)h(\theta - s)\, ds = \frac{1}{2\pi} \int_{-\pi}^{\pi} g(\theta - s)h(s)\, ds$$

which is symmetric in the two functions. Let us calculate the Fourier coefficients of this convolution. We have (using temporarily the notation a_k for the Fourier coefficients)

$$
\begin{aligned}
a_k &= \frac{1}{2\pi} \int_{-\pi}^{\pi} m(\theta)e^{-ik\theta}\, d\theta \\
&= \frac{1}{(2\pi)^2} \int_{-\pi}^{\pi} \int_{-\pi}^{\pi} g(s)h(\theta - s)e^{-ik\theta}e^{iks}e^{-iks}\, ds\, d\theta \\
&= \frac{1}{2\pi} \int_{-\pi}^{\pi} g(s)e^{-iks}\, ds\, \frac{1}{2\pi} \int_{-\pi}^{\pi} h(\theta - s)e^{-ik(\theta - s)}\, d\theta
\end{aligned}
$$

Since $h(\theta - s)$ is assumed to be periodic, we can shift the range of integration by shifting from the variable θ to $u = \theta - s$, and we can see that the second integral is now d_k, while the first integral, since d_k does not depend on s, is c_k. Thus we have shown that

$$a_k = c_k d_k$$

as we set out to do. What we now have is the general statement of what we already knew: due to linearity, the effect of convolving a function with another function merely multiplies the coefficients of the corresponding Fourier expansions.

If in the convolution of two sequences

$$\sum_{k=-\infty}^{\infty} c_k d_{n-k}$$

we suppose that all the d_k are zero except one, say $d_0 = 1$, then the sum

becomes one term, c_n. As we calculate the successive values of the convolution of d_k with c_k, the terms of c_n emerge one at a time. Thus a single
impulse, that is, a selected set of d_k having only one nonzero term, gives
the *impulse response* that is simply the terms of the other sequence.

In the z-transform notation the impulse is often given a special notation
in the form of the *Dirac delta function*

$$\delta(k) = \begin{cases} 1 & k = 0 \\ 0 & k \neq 0 \end{cases}$$

From this it follows easily that

$$\delta(n - k) = \delta(k - n) = \begin{cases} 1 & n = k \\ 0 & n \neq k \end{cases}$$

The delta function can be used to pick out the nth term of a sequence since

$$\sum_{k=-\infty}^{\infty} c_k \delta(n - k) = c_n$$

The sum of the delta function is the unit step function, often labeled
$U(n)$ and defined by

$$U(k) = \sum_{k=-\infty}^{\infty} \delta(k) = \begin{cases} 1 & k \geq 0 \\ 0 & k < 0 \end{cases}$$

The unit step function can be used to express a rectangular window
running from $-N$ to N (or any other range you wish) by writing

$$U(k + N) - U(k - N)$$

In its turn the $U(n)$ function may be summed to get the ramp function
which rises linearly. Combinations of these functions may be used to express
any curve consisting of broken straight lines.

Returning to our definition of a nonrecursive filter

$$y_n = \sum_{k=-\infty}^{\infty} c_k u_{n-k}$$

if we use the impulse function $u_k = 0$ except for $u_0 = 1$, we will obtain
values for the y_n that are simply the coefficients of the filter, c_n, in order;
the impulse produces a response that gives the coefficients of the filter.

Conversely, if we know the impulse response of a nonrecursive filter,
then we know the coefficients and hence the filter. Thus the impulse response

defines the filter and plays a fundamental role in the theory. But we do not emphasize it in this book.

Exercises

5.2-1   Multiply the power series

$$g(x) = \sum_{k=0}^{\infty} a_k x^k \quad \text{and} \quad h(x) = \sum_{m=0}^{\infty} b_m x^m$$

and compare with the convolution.

5.2-2   Using the $U(n)$ function write a rectangular function from $k = a$ to $k = b$.

5.2-3   If $R(k) = \{{}_{k+1}^{0} \quad {}_{k \geq 0}^{k \leq 0}$ is the ramp function, write the formula for an isosceles triangle starting at $-a$, rising to height b at 0, and falling to zero at a.

5.3   THE GIBBS PHENOMENON

The purpose of this section is to investigate the Gibbs phenomenon, named after J. Willard Gibbs (who first publicized this effect, but was not the first to publish it). The phenomenon is so important to the theory of filters that we will first give the treatment in terms of the real sines and cosines, and later give the complex form of the material.

The first example of Section 4.3, which was the expansion of the rectangular pulse function, is typical of a discontinuous function. This function was defined as

$$g(\theta) = \begin{cases} -\frac{1}{2}, & -\pi < \theta < 0 \\ \frac{1}{2} & 0 < \theta < \pi \end{cases}$$

and has a discontinuity of size 1 at $\theta = 0$. The formal Fourier expansion was found to be (Section 4.3, Example 1)

$$g(\theta) = \frac{2}{\pi} \sum_{k=0}^{\infty} \frac{\sin(2k + 1)\theta}{2k + 1}$$

We can write the truncated series of partial sums of $g(\theta)$ as

$$g_N(\theta) = \frac{2}{\pi} \sum_{k=0}^{N} \left[\int_0^{\theta} \cos(2k + 1)u \, du \right] = \frac{2}{\pi} \int_0^{\theta} \left[\sum_{k=0}^{N} \cos(2k + 1)u \right] du$$

To get the sum of the cosines s_N, we multiply, as before, by the sine of the angle of one-half the difference of the angles in successive terms; the result is (compare with Section 4.7)

$$s_N \sin u = \sum_{k=0}^{N} \sin u \cos(2k + 1)u = \frac{1}{2} \sin 2(N + 1)u$$

Thus we have the partial sums as integral

$$g_N(\theta) = \frac{1}{\pi} \int_0^\theta \frac{\sin 2(N + 1)u}{\sin u} \, du$$

Figure 5.3-1 illustrates the partial sum behavior as a function of θ in the case of five terms and with the independent variable $x = \theta/\pi$. We see an overshoot of the curve of the partial sums. To find where the peaks and valleys occur, we differentiate the function and set it equal to zero. This procedure is the same as equating the integrand to zero. Clearly, the first maximum occurs at $\theta = \pi/2(N + 1)$. The value of the partial sum at this point is

$$g_N \left[\frac{\pi}{2(N + 1)} \right] = \frac{1}{\pi} \int_0^{\pi/2(N+1)} \frac{\sin 2(N + 1)u}{\sin u} \, du$$

Our problem is to estimate this value as N approaches infinity. Using

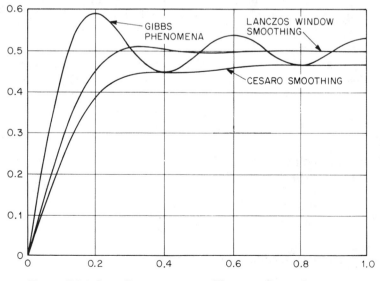

FIGURE 5.3-1 GIBBS PHENOMENA WITH WINDOWS: PLAIN, LANCZOS, AND CESARO WINDOWS

the change of the dummy variable of integration $v = 2(N + 1)u$ gives (when we insert extra v terms)

$$g_N \left[\frac{\pi}{2(N + 1)} \right] = \frac{1}{\pi} \int_0^\pi \left(\frac{\sin v}{v} \right) \left[\frac{\{v/2(N + 1)\}}{\sin\{v/2(N + 1)\}} \right] dv$$

Obviously, since $(\sin \theta)/\theta$ approaches 1 as θ approaches 0, the square bracket approaches 1 as N gets larger and larger. Thus in the limit we have, for the value of the first overshoot,

$$g_\infty(0) = \frac{1}{\pi} \int_0^\pi \frac{\sin v}{v} dv$$

This is the standard sine integral function. From tables, we get the value for the first maximum

$$\frac{1.17898}{2} = 0.58949 = 0.5 + 0.08949$$

The next local minimum has the corresponding limiting value of

$$0.45142 = 0.5 - 0.04858$$

These values lead (when N is large) to the values of approximately 9% overshoot and 5% undershoot for the unit jump (discontinuity) in the function.

   This Gibbs phenomenon occurs whenever we truncate a Fourier series. It will be of great importance when discussing the design of filters because, for practical purposes, we are forced to truncate any Fourier series we generate.

5.4 LANCZOS SMOOTHING: THE SIGMA FACTORS

   Cornelius Lanczos observed that the ripple in the sum of the truncated series has the period of either the first term neglected or the last term kept. In either case, he argued, smoothing the partial sum by integrating (averaging) over this period would remove the main effects of the ripple.

   We apply this idea to the general truncated Fourier series

$$g_N(\theta) = \frac{a_0}{2} + \sum_{k=1}^N [a_k \cos k\theta + b_k \sin k\theta]$$

For the smoothed value, we take the *average* over the interval of length $2\pi/N$ centered at θ. (Here we pick N as the order of the last term kept so that the reader can watch the last term disappear; but, in practice, we pick N as the order of the first term neglected.) We have, therefore, the smoothed value $h_N(\theta)$ as the average of the $g_N(\theta)$.

$$h_N(\theta) = \frac{N}{2\pi} \int_{\theta - (\pi/N)}^{\theta + (\pi/N)} g_N(s)\, ds$$

Working out the details, we have

$$h_N(\theta) = \left(\frac{a_0}{2}\right)\left(\frac{N}{2\pi}\right) \int_{\theta - (\pi/N)}^{\theta + (\pi/N)} ds$$
$$+ \frac{N}{2\pi} \sum_{k=1}^{N} \left[a_k \int_{\theta - (\pi/N)}^{\theta + (\pi/N)} \cos ks\, ds + b_k \int_{\theta - (\pi/N)}^{\theta + (\pi/N)} \sin ks\, ds \right]$$

$$h_N(\theta) = \frac{a_0}{2} + \frac{N}{2\pi} \sum_{k=1}^{N} \left[a_k \left\{ \frac{\sin k[\theta + (\pi/N)] - \sin k[\theta - (\pi/N)]}{k} \right\} \right.$$
$$\left. - b_k \left\{ \frac{\cos k[\theta + (\pi/N)] - \cos k[\theta - (\pi/N)]}{k} \right\} \right]$$

Using the trigonometric formula for the difference of two sines and the formula for the difference of two cosines, we finally obtain

$$h_N(\theta) = \frac{a_0}{2} + \sum_{k=1}^{N} \sigma(N, k)[a_k \cos k\theta + b_k \sin k\theta]$$

where $\sigma(N, k)$ are the *sigma factors*:

$$\sigma(N, k) = \frac{\sin \pi k/N}{\pi k/N}$$

Thus the smoothed Fourier series is the original Fourier series with its coefficients multiplied by the corresponding sigma factors.

This result is not surprising; since the smoothing operation was linear, some kind of multiplying factor had to appear. When $k = N$, the sigma factor is zero, and it does not matter whether the last term kept is called N or the first term neglected is called N: the effect is that the term of frequency N has the multiplier equal to zero.

The formula for $h_N(\theta)$ is the average of $g_N(\theta)$ for a symmetric interval about θ and of width $2\pi/N$. We may view this smoothing operation as if looking at the original function $g(\theta)$ through a narrow, translucent, rectangular window of width $2\pi/N$. The function $h_N(\theta)$ that we see is the average light intensity from the original function $g_N(\theta)$, which is imagined as being brighter when the function $g_N(\theta)$ is higher and less bright when the function is smaller. It is easily seen that the resultant function $h_N(\theta)$, the Lanczos smoothed curve, has the high-frequency ripples somewhat smoothed out, as shown by Figure 5.3-1. The first overshoot is reduced from 0.08949 to 0.01187 (about $\frac{1}{8}$ as much) and the first minimum from 0.04858 to 0.00473 (about $\frac{1}{10}$ as much).

In selecting our smoothing interval, we adjusted it to the number of terms being kept in the Fourier series; that is, we had the same N in both the number of terms and in the smoothing interval or, what is the same thing, in the sigma factors. An examination of the process of deriving the sigma factors shows that, if we do not make these two the same (or differing by 1), then we will still obtain sigma factors that are related to the smoothing interval selected. A study of the sigma factors shows that in this situation the series will not necessarily end with the Nth sigma factor that has the value 0 but can continue with additional sigma factors that are not zero. Viewed as a function of k, the sigma factors when k is greater than N have a decaying ripple of period $2N$ and are slowly alternating in sign.

The theory of Fourier series has another smoothing formula that is widely used (mainly in mathematical circles), the averaging of the successive partial sums $g_N(\theta)$. The process is called *Fejer smoothing* (also called Cesaro 1). Fejer smoothing produces the weighting of the coefficients of the series by

$$c(N, k) = \frac{N - k}{N}$$

Thus we have the Fejer smoothed series

$$\text{Fejer } g_N(\theta) = \frac{a_0}{2} + \sum_{k=1}^{N} \frac{N - k}{N} [a_k \cos k\theta + b_k \sin k\theta]$$

Fejer smoothing of the rectangular pulse function is also shown in Fig. 5.3-1 and illustrates the point that the "rise time" of Fejer smoothing is very much longer than for Lanczos smoothing. Consequently, Fejer smoothing is seldom used in practice. It is chiefly mathematicians who use it. In the limit, as N approaches infinity, this curve has nice mathematical properties; for finite values, however, it approaches its final value much too slowly.

We have examined one particular function with a jump, but it is typical of functions with jumps in them. We can expect the Gibbs phenomenon whenever a function has a discontinuity, and we can also expect the smoothing formulas to produce their corresponding effects.

It is much easier to derive the sigma factors in the complex form. If we take the truncated series in the form

$$g_N(\theta) = \sum_{k=-N}^{N} c_k e^{ik\theta}$$

then the Lanczos smoothing derivation goes as follows:

$$h_N(\theta) = \frac{N}{2\pi} \sum_{k=-N}^{N} c_k \int_{\theta-\pi/N}^{\theta+\pi/N} e^{ik\theta} \, d\theta$$

$$= \sum_{k=-N}^{N} c_k \left(\frac{N}{2\pi}\right) e^{ik\theta} \left[\frac{e^{i(\pi k/N)} - e^{-i(\pi k/N)}}{ik}\right]$$

$$= \sum_{k=-N}^{N} c_k \left[\frac{\sin k\pi/N}{\pi k/N}\right] e^{ik\theta}$$

This neat, compact derivation, when compared to the earlier "trigonometric" derivation, should convince you that the complex exponentials not only are easier to handle but also add clarity to what is going on.

5.5 THE GIBBS PHENOMENON AGAIN

Using the second form of the convolution theorem, Section 5.2, we can now see the Gibbs phenomenon in an entirely different light. Given a function represented as a Fourier series

$$g(\theta) = \sum_{k=-\infty}^{\infty} c_k e^{ik\theta}$$

the act of truncating this series to

$$g_N(\theta) = \sum_{k=-N}^{k=N} c_k e^{ik\theta} \tag{5.5-1}$$

is the same as multiplying the coefficients c_k by the numbers ($2N + 1$ values each equal to 1)

$$0, 0, 0, 1, 1, \ldots, 1, 0, 0, 0, \ldots$$

or

$$d_k = \begin{cases} 1, & |k| \le N \\ 0, & |k| > N \end{cases} \tag{5.5-2}$$

What function $h(\theta)$ has these $2N + 1$ nonzero coefficients d_k? Clearly,

$$h(\theta) = \sum_{k=-N}^{k=N} e^{ik\theta} = e^{-iN\theta} + e^{-i(N-1)\theta} + \cdots + e^{iN\theta} \qquad (5.5\text{-}3)$$

In Section 1.8 we summed this geometric progression:

$$h(\theta) = \frac{e^{i(N+1/2)\theta} - e^{-i(N+1/2)\theta}}{e^{i\theta/2} - e^{-i\theta/2}}$$

$$= \frac{\sin(N + 1/2)\theta}{\sin(\theta/2)} \qquad (\,|\,\theta\,| < \pi) \qquad (5.5\text{-}4)$$

It is customary to *normalize* a window so that it has unit area, meaning in this case that the sum of the coefficients should total to 1. Thus the window 5.5-2 becomes

$$d_k = \begin{cases} \dfrac{1}{2N + 1}, & |\,k\,| \le N \\[2ex] 0 & |\,k\,| > N \end{cases} \qquad (5.5\text{-}5)$$

and 5.5-4 becomes correspondingly

$$h(\theta) = \frac{\sin(N + 1/2)\theta}{(2N + 1)\sin \theta/2} \qquad (5.5\text{-}6)$$

This is the corresponding (convolving) window to 5.5-5. For plotting purposes it is convenient to change to rotations,

$$\theta = 2\pi f$$

For large N, this is a rapidly oscillating function with a maximum value of 1 at $\theta = 0$ (L'Hôpital's rule will show this), and falling off in amplitude as the denominator grows larger. See Figure 5.5-1 for the case $N = 5$ (unmodified rectangular window).

Thus, corresponding to the discrete rectangular window (5.5-5), we have the rapidly oscillating continuous window (5.5-6). When this window is convolved with a discontinuous square cornered function, as in Figure 5.5-2, we see how the integral of the product of the two functions produces the wiggles of the Gibbs phenomenon as a function of the displacement between the centers of the functions. Thus we see how the Gibbs phenomenon arises in a completely different light than in Section 5.3.

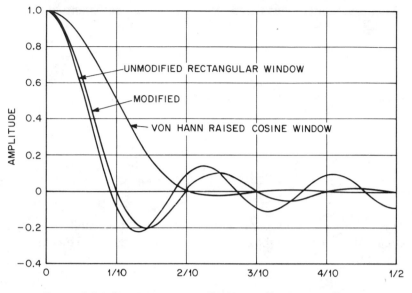

FIGURE 5.5-1 COMPARISON OF THE FREQUENCY RESPONSE OF THREE WINDOW FUNCTIONS, 5 TERMS

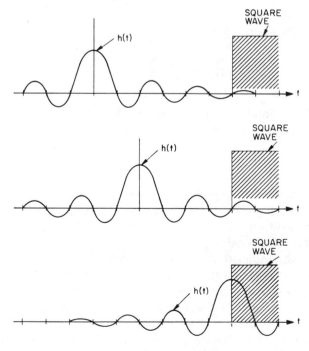

FIGURE 5.5-2 GIBBS PHENOMENON

5.6 MODIFIED FOURIER SERIES

Instead of truncating a Fourier series by multiplying its coefficients by

$$\ldots 0, 0, 0, 1, 1, 1, \ldots , 1, 1, 0, 0, 0, \ldots$$

we can use the sequence having $\frac{1}{2}$ as the two end values:

$$\ldots 0, 0, 0, \tfrac{1}{2}, 1, 1, \ldots , 1, \tfrac{1}{2}, 0, 0, 0, \ldots \tag{5.6-1}$$

(see section 3.4). For this *modified rectangular window*, we get not the earlier $h(\theta)$ but one that is decreased by

$$\tfrac{1}{2}(e^{iN\theta} + e^{-iN\theta}) = \cos N\theta$$

So our new $h(\theta)$ is

$$h(0) - \frac{\sin(N + 1/2)\theta}{\sin \theta/2} - \cos N\theta$$

Expanding the first term, we get

$$h(\theta) = \frac{\sin N\theta \cos \theta/2 + \cos N\theta \sin \theta/2 - \cos N\theta \sin \theta/2}{\sin \theta/2}$$

or

$$h(\theta) = \frac{\sin N\theta \cos \theta/2}{\sin \theta/2} \tag{5.6-2}$$

If we again normalize this window to have unit area, which is the same as setting the sum of the weights equal to 1, we must divide each weight, and hence the result, by $2N$. We have, therefore, corresponding to (5.5-6),

$$h(\theta) = \frac{\sin N\theta \cos \theta/2}{2N \sin \theta/2} \tag{5.6-3}$$

Comparing this (Figure 5.5-1) with the earlier unmodified rectangular window, we see a slight change in the high-frequency term but also an extra factor of $\cos \theta/2$. This extra factor begins with the value of 1 at $\theta = 0$ and decreases to 0 at $\theta = \pi$ (the folding frequency). Consequently, this con- volving window is significantly smaller at the ends and thus produces "slightly better results" in many situations. However, working out the corresponding overshoot for $N \to \infty$, it turns out to be the same; the windows are almost the same at the center but differ significantly near the ends.

As a result of the theory of convergence of a Fourier series at a point of discontinuity, it is not surprising that this modified rectangular window can give better results. But it does raise a central point, one that will be discussed in detail in the future: how to deal with a finite piece of an infinite record. The use of a rectangular window clearly has bad properties, whereas even the simplest modification produces some improvement.

5.7 THE VON HANN WINDOW: THE RAISED COSINE WINDOW

Truncating an array of Fourier series coefficients is exactly the same as looking at the original function through a convolving window. Ideally, we would like a tall, narrow window of unit area so that what we "see" is very close to the original function at the middle of the (symmetric) window. Unfortunately, we also want to use as few terms as possible in the Fourier series, which means using a wide window.

One way the width of a window may be measured is by the distance between the closest zeros on each side of the center lobe, which is the main lobe of the window. But it can also be measured in a different sense by the *size* of the lobes that are farther away from the center. We have seen (Section 5.5) that the main lobe gives rise to the transition band (a region between a stopband and a passband), and the side lobes give rise to the wiggles (the Gibbs phenomenon), which may be regarded as "contamination," or "leakage," from parts of the function that are adjacent to where we are looking.

Because we are interested in transfer functions (although the results apply to arbitrary functions), we shift to the f notation.

The simple rectangular window (equation 5.5-2 with $\theta = 2\pi f$) gives rise to the convolving window through which we see the function (5.5-6).

$$h(f) = \frac{\sin[\pi(2N + 1)f]}{\sin \pi f}$$

The modified rectangular window (equation 5.6-1) merely takes one-half of the end values, but it moves the first zeros out from

$$f = \pm \frac{1}{2N + 1} \quad \text{to} \quad f = \pm \frac{1}{2N}$$

It also puts in an extra cosine factor that steadily decreases the sizes of the side lobes that are farther and farther away from the center of the window. For $N = 5$ (11 terms in the final filter), Figure 5.5-1 illustrates this change; the slight loss in narrowness is more than compensated for by the decrease in the heights of the other lobes. The sigma factors from the Lanczos window also show how tapering the weights applied to the coefficients of the Fourier series can greatly decrease the heights of the side lobes. The third curve in the figure will be examined next.

The foregoing remarks suggest examining a more severe weighting of the Fourier coefficients which we keep in the truncation process. One example is the *von Hann window* which is also called the *raised cosine window* from its definition:

$$w_k = \begin{cases} \dfrac{1 + \cos \pi k/N}{2}, & |k| \le N \\[2mm] 0, & |k| \ge N \end{cases} \tag{5.7-1}$$

The weighting sequence of $2N + 1$ terms, when viewed as a continuous function, not only vanishes at the ends but is also tangent there.

The corresponding transform in the frequency domain of the von Hann window is, by definition, the function

$$\tilde{h}(f) = \sum_{k=-N}^{k=N} w_k e^{2\pi i k f}$$

$$= \tfrac{1}{2} \sum_{k=-N}^{k=N} \left(1 + \frac{e^{\pi i k/N} + e^{-\pi i k/N}}{2} \right) e^{2\pi i k f}$$

$$= \tfrac{1}{4} \sum_{k=-N}^{k=N} (e^{i\pi k/N} + 2 + e^{-i\pi k/N}) e^{2\pi i k f}$$

Since $w_N = 0$, taking half the end values has no effect, and the results of Section 5.5 can be applied to the three separate terms in the summation. This step gives

$$\tilde{h}(f) = \frac{1}{4} \left\{ \frac{\sin[\pi(f + 1/2N)2N] \cos[\pi(f + 1/2N)]}{\sin[\pi(f + 1/2N)]} + 2 \frac{\sin \pi f 2N \cos \pi f}{\sin \pi f} \right.$$

$$\left. + \frac{\sin[\pi(f - 1/2N)2N] \cos[\pi(f - 1/2N)]}{\sin \pi(f - 1/2N)} \right\}$$

which may be written as

$$\tilde{h}(f) = \frac{1}{4} \left\{ \frac{\sin \pi(2Nf + 1) \cos[\pi(f + 1/2N)]}{\sin[\pi(f + 1/2N)]} + \frac{2 \sin \pi f 2N \cos \pi f}{\sin \pi f} \right.$$

$$\left. + \frac{\sin \pi(2Nf - 1) \cos[\pi(f - 1/2N)]}{\sin[\pi(f - 1/2N)]} \right\}$$

Expanding the sines in the numerators of the end terms, we get ($\sin \pi = 0$ and $\cos \pi = -1$)

$$h(f) = \frac{\sin 2\pi N f}{4} \left\{ -\cot \pi\left(f + \frac{1}{2N}\right) + 2 \cot \pi f - \cot \pi\left(f - \frac{1}{2N}\right) \right\}$$

Next, applying the trigonometric identity

$$\cot(a + b) + \cot(a - b) = \frac{2 \sin a \cos a}{\sin^2 a - \sin^2 b}$$

gives

$$h(f) = \frac{\sin 2\pi N f}{4} \left\{ \frac{2 \cos \pi f}{\sin \pi f} - \frac{2 \sin \pi f \cos \pi f}{\sin^2 \pi f - \sin^2(\pi/2N)} \right\}$$

This result may be rewritten in a form that shows the dependence on the parameters as follows:

$$h(f) = \frac{\sin 2\pi N f \cos \pi f}{2 \sin \pi f} \left[\frac{1}{1 - \left(\dfrac{\sin \pi f}{\sin \pi/2N}\right)^2} \right] \tag{5.7-2}$$

The zeros of the denominator in the bracket ($f = \pm 1/2N$) are compensated by corresponding zeros in the front multiplier, and $h(f)$ has the finite value $N/2$ at $f = \pm 1/(2N)$. As $f \to 0$,

$$h(0) \longrightarrow \frac{2\pi N}{2\pi} = N$$

The first zeros of $h(f)$ occur at $f = 1/N$. Thus the window is twice as wide as the modified window, *but* the side lobes are greatly reduced, as shown in Figure 5.5-1. In a sense, this window is the sum of three modified windows, one at $f = -1/2N$ of size $\frac{1}{4}$, one at $f = 0$ of size $\frac{1}{2}$, and, finally, one at $f = 1/2N$ of size $\frac{1}{4}$.

5.8 HAMMING WINDOW: RAISED COSINE WITH A PLATFORM

From Figure 5.5-1 it is evident that the von Hann window and the modified window have opposite signs in the side lobes. This situation immediately suggests that a small amount of the modified window added to the von Hann window could be used to reduce the maximum that occurs in the side lobes. The outcome is the *Hamming window* (sometimes written as "hamming window" [BT, p. 98]), which has coefficients 0.23, 0.54, 0.23, instead of the von Hann coefficients 0.25, 0.50, 0.25.

To find the coefficients of the Hamming window, we merely take a weighted sum of the modified rectangular window and the von Hann window and use an optimizing routine (see Section 9.10) to find the weights that minimize the maximum of the side lobes (tails) of the window. This is the Chebyshev criterion, which will be discussed in Chapter 12. The value obtained for the weights depends, of course, on the value of N used in the weighting window $\{w_k\}$. As can be seen from Figure 5.8-1, the values of a and b in the Fourier expansion vary slowly with $1/N$ for reasonably long runs of data. Notice that $2a + b = 1$ for all values of $1/N$. The result of this optimization appears in Figure 5.8-2, which gives the small corrections to the Hamming window in terms of a quantity labeled d. The reason for the peculiar shape is revealed by a detailed look at the tails of the window.

Because of the central role of least squares, the idea of minimizing the integral of the square of the side lobes *relative to* the integral of the square of the main lobe will probably occur to most people. Both the Chebyshev minimum maximum error and the least-squares error are given in Figure 5.8-3, where we see the large negative first lobe of the least-squares optimization.

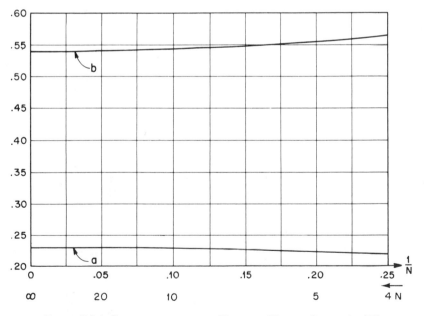

Figure 5.8-1 Coefficients for the Hamming Window $2a \cos (\pi n/N)$ $+ b$

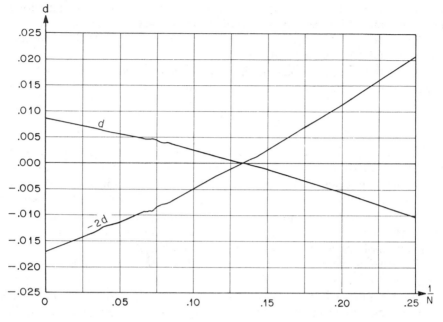

FIGURE 5.8-2 COEFFICIENTS FOR THE OPTIMAL LEAST SQUARES WINDOW, $a = 0.23 + d$, $b = 0.54 - 2d$

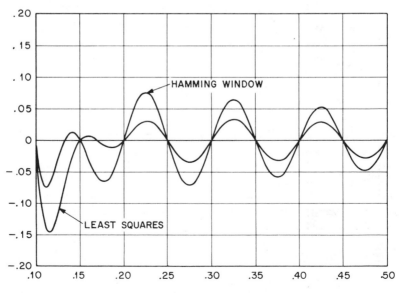

FIGURE 5.8-3 FREQUENCY RESPONSE

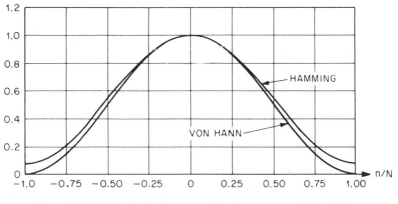

FIGURE 5.8-4 WEIGHT FACTORS FOR HAMMING AND VON HANN WINDOWS

The general shape of the weighting factors of the von Hann and Hamming windows is illustrated in Figure 5.8-4. The figure shows why the Hamming window is often referred to "as a raised cosine on a platform."

Exercises

5.8-1 For the Blackman window

$$w_k = 0.42 + 0.50 \cos \frac{\pi k}{N} + 0.08 \cos \frac{2\pi k}{N}$$

find the $W(f)$. Plot $W(f)$ and compare with Hamming and von Hann windows.

5.8-2 Do the same for triangular windows.

5.9 REVIEW OF WINDOWS

Because windows are the most confusing part of digital filter theory, let us review the situation.

We begin with a continuous signal $u(t)$ and sample it at unit intervals in order to obtain the set of measurements $\{u_n\}$. When we limit the signal to the range $-N \le n \le N$, it follows from the second convolution theorem (Section 5.2) that the truncation process is equivalent to "smearing the spectrum of the signal" by effectively looking at the true spectrum through the translucent convolving window (equation 5.5-6)

$$h(\omega) = \frac{\sin(N + 1/2)\omega}{(2N + 1) \sin \omega/2}$$

If instead of merely looking at a section of the sequence of data points $\{u_n\}$, we also weight it with weights $\{w_n\}$; that is, we use the sequence $\{w_n u_n\}$ in place of $\{u_n\}$, then we get a different window. Merely taking half the end values (5.6-1) changes the convolving window to (5.6-3):

$$h(\omega) = \frac{\sin N\omega \cos \omega/2}{2N \sin \omega/2}$$

Although doing so somewhat decreases the side lobes, they are still large enough to produce significant distortions of the original transform.

Further modification of the weights leads to the von Hann window (Section 5.7) with significantly smaller side lobes, but also with twice the width in the main lobe (which means a loss in distinguishing close details). A slight additional modification leads to the Hamming window (Section 5.8), which has the smallest extreme value in the side lobes.

However, we can also use the first convolution theorem. Thus, if we use one of the sequences

$$\{\tfrac{1}{4}, \tfrac{1}{2}, \tfrac{1}{4}\} \qquad \text{von Hann}$$

$$\{0.23, 0.54, 0.23\} \qquad \text{Hamming}$$

as a smoothing formula on the original data $\{u_n\}$, we effectively multiply the transform by the corresponding continuous windows (Figure 5.8-4). These windows tend to remove the higher frequencies.

There are many types of windows that may be used in either of the two ways, on the function $u(t)$ or on the transfer function $H(\omega)$, and an elementary textbook is not the place to discuss all of them (see IEEE-I).

Appendix 5.A

A sequence S_N is said to converge to S ($S_N \rightarrow S$) if no matter how close one asks (given an $\epsilon > 0$) it is possible to find a place N_0 (there exists an N_0) such that for all greater values of N (for all $N \geq N_0$) the values of S_N are all closer to S than required ($|S_N - S| < \epsilon$).

The convergence of a series is reduced to the convergence of a sequence by the simple device of considering the sequence of partial sums of the series.

If we now think of a series whose terms depend on a variable, say θ, then for each θ we can consider the convergence. In this situation, the N_0 will depend on *both* ϵ and θ,

$$N_0 = N_0(\epsilon, \theta)$$

In a given interval (open or closed), it may be possible to find an N_0 *independent* of θ that will meet the convergence condition for a given ϵ. If so, the series is said to be *uniformly convergent*; otherwise, the series is not uniformly convergent.

6

Design of Nonrecursive Filters

6.1 INTRODUCTION

Now that we have the necessary mathematical theory, we are ready to design nonrecursive digital filters. Recall that the typical smoothing filter was a *low-pass filter*, meaning that the low frequencies "pass" through and the high frequencies are "stopped" (eliminated), with a transition zone between the passband and stopband (of frequencies). (See Figure 6.1-1.) In Chapter 3 we designed such filters by choosing a finite, symmetric set of coefficients. We selected a digital filter of the form

$$y_n = \sum_{k=-N}^{k=N} c_k u_{n-k} \qquad (c_k = c_{-k}) \tag{6.1-1}$$

Had we wanted to interpolate missing data, we would have picked much the same type of digital filter but set $c_0 = 0$.

A *high-pass filter* is, of course, the opposite of a low-pass filter: it passes the high frequencies and stops the low (Figure 6.1-2). It is simply the difference between an "all-pass filter" $y_n = u_n$ and a low-pass filter. That is, from the low-pass filter having coefficients c_k we get the high-pass filter (primed coefficients)

$$\begin{cases} c_k' = -c_k & k \neq 0 \\ c_0' = 1 - c_0 & k = 0 \end{cases}$$

There is a second way of obtaining a high-pass filter from a low-pass

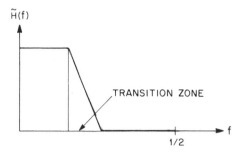

FIGURE 6.1-1 LOW-PASS FILTER

filter. If we replace f by $\frac{1}{2} - f$, we simply reverse the function in the interval $0 \le f \le \frac{1}{2}$. The effect on each term of the cosine series is

$$\cos 2\pi k(\tfrac{1}{2} - f) = \cos \pi k \cos 2\pi kf$$

$$= (-1)^k \cos 2\pi kf \qquad (6.1\text{-}2)$$

The effect, therefore, is merely to change the sign of the odd harmonics of the filter. Since the cosine is an even function, the corresponding change also occurs on the negative half of the frequency interval. Both methods can be combined in tandem to get a high-pass filter from a low-pass filter.

There are also *bandpass* and *bandstop* filters. A bandpass filter is often used to study a part of a spectrum. Frequently, a very narrow bandstop filter is called a *notch filter*. Among other uses, a notch filter (Figure 6.1-3) is used to remove the ever present 60 Hz that comes from our electrical power distribution system (in the United States; in England it is 50 Hz).

Differentiation filters require odd symmetry in the coefficients (and thus lead to only sine terms in the Fourier expansion); they have $c_k = -c_{-k}$ and $c_0 = 0$. Integration cannot be done by nonrecursive filters.

In Section 4.4, we showed that any function can be written as the sum of an even and an odd function. Similarly, the identity

$$c_k = \frac{c_k + c_{-k}}{2} + \frac{c_k - c_{-k}}{2} \qquad (6.1\text{-}3)$$

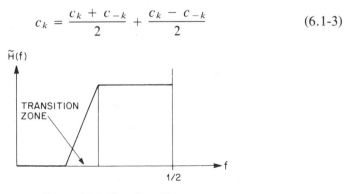

FIGURE 6.1-2 HIGH-PASS FILTER

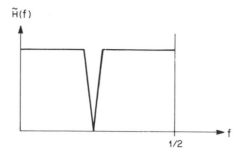

FIGURE 6.1-3 NOTCH FILTER

shows that any nonrecursive digital filter can be written as the sum of a smoothing (even) filter and a differentiating (odd) filter. A smoothing filter can be viewed as a linear combination of the sums of symmetrically placed data, while a differentiating filter uses differences. Clearly, they are simply the cosine and sine terms of the general Fourier series expansion.

Given a symmetric digital filter, $c_k = c_{-k}$, we substitute $u_n = e^{i\omega n} = z^n$ into the equation and obtain the transfer function $H(\omega)$. As a result of the symmetry of the coefficients, we have a Fourier series in cosines and a Nyquist interval of $-\pi \leq \omega \leq \pi$. In the frequency notation f, the Nyquist interval, is $-\frac{1}{2} \leq f \leq \frac{1}{2}$. The transfer function $\tilde{H}(f)$ is symmetric about $f = 0$. A graph of the values of this Fourier series gives the curve of the transfer function.

We can reverse these steps (Figure 6.1-4). Given an arbitrarily shaped (symmetric) transfer function [part (a) of Figure 6.1-4], we can find, using the methods of Sections 4.4 and 4.9, the coefficients of the corresponding cosine Fourier series [part (b)]. This Fourier series will, in general, have an infinite number of coefficients; in practice, we want a finite filter, and so we are forced to truncate the coefficients past some subscript value N [part (c)]. But the truncation produces the Gibbs phenomenon, as discussed in Sections 5.3 and 5.5 [part (d)]. To avoid it, we apply the rectangular convolving window (Section 5.4), which multiplies the coefficients by the sigma factors [part (e)]. This process, in turn, produces the smoothed transfer function [part (f)]. The total effect on the coefficients c_k of the complex Fourier series expansion of the transfer function is that we multiply the c_k by the sigma factors $\sigma(N, k)$ for $|k| \leq N$ and by 0 for $|k| > N$. The sigma factors are, therefore, the coefficient multiplying window. In the transform, this window is the combination of first creating the Gibbs phenomenon and then doing the Lanczos smoothing.

The use of the von Hann or Hamming windows in place of the rectangular window Lanczos used greatly reduces the ripples in the final transfer function but also doubles the width of the transition band.

In our basic design method (illustrated in Figure 6.1-4), the step going to part (b) finds the Fourier coefficients that are the least-squares fit (Section

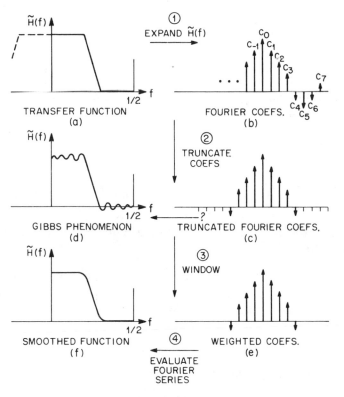

FIGURE 6.1-4

4.4). The sigma factors alter this basic least-squares fit. In the next section the method will be illustrated by a specific example.

To summarize our first method of design, given a symmetric transfer function $H(\omega)$, we perform the following steps: find the first N Fourier cosine coefficients $a_0, a_1, \ldots, a_{N-1}$; then multiply these coefficients by the corresponding sigma factors, using the same value N; convert the resulting coefficients to the appropriate c_k (of the filter) by dividing by 2 (beware of the constant term); and, finally, plot the resulting transfer function to check the results.

6.2 A LOW-PASS FILTER DESIGN

We will now design a low-pass filter giving some numerical details. Our basic design method is that of Figure 6.1-4. As a first, concrete example, we select the transfer function [Figure 6.2-1, curve (a)]

$$\tilde{H}(f) = \begin{cases} 1, & 0 < |f| < 0.2 \\ 0, & 0.2 < |f| < 0.5 \end{cases}$$

with, of course,

$$\check{H}(-f) = \check{H}(f)$$

We are at (a) of Figure 6.1-4.

By direct computation of the integrals for the coefficients of the Fourier series (Sections 4.3 and 4.4), all the $b_k = 0$, and

$$a_k = \frac{2}{\pi} \int_0^\pi H(\omega) \cos k\omega \, d\omega$$

$$= 4 \int_0^{1/2} \check{H}(f) \cos 2\pi k f \, df$$

Thus we have for the coefficients of our particular ideal transfer function of a low-pass filter

$$a_k = 4 \int_0^{0.2} \cos 2\pi k f \, df = \frac{2}{\pi k} \sin 0.4\pi k = \begin{cases} k = 0 & 0.80000 \\ k = 1 & 0.30273 \\ k = 2 & 0.09355 \\ k = 3 & -0.06237 \\ k = 4 & -0.07568 \end{cases}$$

and the corresponding Fourier series [Figure 6.1-4(b)] is

$$\check{H}(f) = \frac{4}{10} + 2 \sum_{k=1}^\infty \left[\frac{\sin(4k\pi/10)}{\pi k} \right] \cos 2\pi k f$$

We are now at (b) of Figure 6.1-4.

For practical reasons, we *truncate the infinite series to a finite length.* We will use $N = 5$ terms, and so we set $N - 1 = 4$ in the summation

$$\check{H}(f) = \frac{4}{10} + 2 \sum_{k=1}^{k=4} \left[\frac{\sin(4k\pi/10)}{\pi k} \right] \cos 2\pi k f$$

We are now at part (c) in Figure 6.1-4 and face the Gibbs phenomenon at part (d). To reduce the size of the ripples, [Figure 6.2-1, curve (d)] we use the rectangular window, which means multiplying the coefficients of the Fourier expansion by the appropriate sigma factors (Section 5.4); in this case,

$$\sigma(5, k) = \frac{\sin(\pi k/5)}{\pi k/5} = \begin{cases} k = 0 & 1.00000 \\ k = 1 & 0.93589 \\ k = 2 & 0.75683 \\ k = 3 & 0.50455 \\ k = 4 & 0.23387 \end{cases}$$

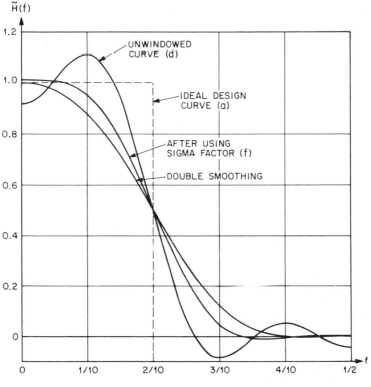

$\tilde{H}(f)$

UNWINDOWED
CURVE (d)

IDEAL DESIGN
CURVE (a)

AFTER USING
SIGMA FACTOR (f)

DOUBLE SMOOTHING

FIGURE 6.2-1 LOW-PASS FILTER, $N = 5$

Notice that $\sigma(5, 5) = 0$ and that we would have removed the term a_5 had we tried to keep it. Thus, for the modified transfer function, we have

$$\tilde{H}(f) = \frac{4}{10} + 2 \sum_{k=1}^{k=4} \left(\frac{\sin(\pi k/5)}{\pi k/5} \right) \left[\frac{\sin(4k\pi/10)}{\pi k} \right] \cos 2\pi kf$$

Having used the rectangular window, we now have the transfer function at part (e) and the Fourier coefficients at part (f) in Figure 6.1-4. Thus we have our filter design as shown in Figure 6.2-1, curve (f) for $N = 5$. See Figure 6.2-2 for the case of $N = 10$. The third curve in each figure will be discussed in the next section. The digital filter coefficients c_k are half those in the cosine expansion except for the constant term.

Exercises

6.2-1   Design a low-pass filter using $N = 5$, $\tilde{H}(f) = 1$ for $|f| < \frac{1}{10}$, and 0 elsewhere.

6.2-2   Design a high-pass filter with $N = 4$ that passes the upper half of the Nyquist interval.

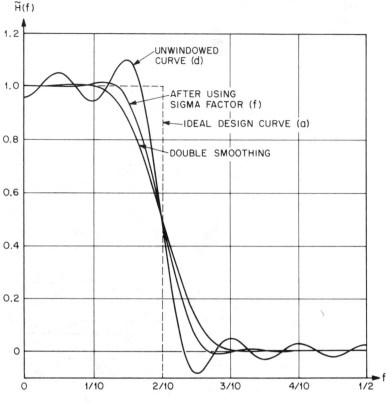

FIGURE 6.2-2 LOW-PASS FILTER, $N = 10$

6.2-3 Design a bandpass filter that passes the middle third of the Nyquist interval (use $N = 6$).

6.2-4 Design a high-pass filter with $\tilde{H}(f) = 1$, $| f | \geq \frac{4}{10}$ and 0 elsewhere (use $N = 5$).

6.2-5 Design a bandpass filter with $\tilde{H}(f) = 1$ for $\frac{2}{10} \leq | f | \leq \frac{3}{10}$ and 0 elsewhere (use $N = 5$).

6.3 CONTINUOUS DESIGN METHODS: A REVIEW

Having designed one specific filter, let us turn, using our earlier developments, to a slightly more general approach to a low-pass filter design.

First, we start with an arbitrary width for the passband; the passband will be from zero to f_s, and the stopband will go from f_s to $\frac{1}{2}$ (measured in

f, rotations). Direct calculation of the Fourier coefficients (remember that this transfer function is an even function of f) gives the expansion

$$\tilde{H}(f) = 2f_s + \sum_{k=1}^{k=\infty} \left[\frac{2}{\pi k} \sin 2\pi k f_s \right] \cos 2\pi k f$$

This expansion agrees with the result of Section 6.2 when $f_s = 0.2$.

After truncating it to a finite number of terms, $2N + 1$, (applying a rectangular multiplying window to the coefficients), we have the Gibbs phenomenon in the frequency domain. A rectangular convolution window is used to smooth this effect. Thus we apply the sigma factors. In Section 6.2 we chose the number of terms $N = 5$ (and $N = 10$) and adjusted the width of the window to the width of the ripples in the Gibbs phenomenon.

Consider, now, selecting the transition width and using it to determine the number of terms. We choose a general rectangular convolving window of width Δ and unit area. Earlier we showed that any linear operation on the function will appear as a multiplication of the Fourier coefficients by some constants, just as it happened in the rectangular window, or in the convolution of two functions. By carrying out the algebra and trigonometry in detail, we will find that, corresponding to the sigma factors, we get the factors

$$\frac{\sin \pi k \Delta}{\pi k \Delta}$$

as multipliers of the corresponding coefficients of the Fourier series. We naturally truncate where these factors first become small.

It is important to note that we could also view this whole process as first smoothing the transfer function. When in our mind's eye we picture the convolution of the rectangular window (dot-dash lines in Figure 6.3-1) going across the original transfer function, we see that (assuming that the window is narrow enough) in the beginning we obtain the same constant

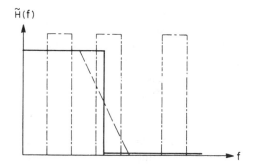

FIGURE 6.3-1 A RECTANGULAR WINDOW ON A DISCONTINUITY

value *until* the window begins to touch the discontinuity (dashed line). As the window crosses the discontinuity, the smoothed value decreases linearly. When the window is completely across the discontinuity, the smoothed value is zero. Thus we may, as an alternate to the original problem of finding an approximation to the discontinuous function, approximate the function with a straight-line interpolation between the two parts, the width of the transition being exactly twice the width of the rectangular window used. (See Figure 6.3-1.) In this way, we see that whether we take the original function and, after truncation, smooth with sigma factors or imagine the rectangular window being convolved with the original function and then fit the resulting smoothed function, we must get the same result (we, of course, truncate the Fourier series in both cases).

The results of the two approaches appear slightly different, although the effect is the same. Either we truncate and smooth or we smooth the curve to get a more rapidly converging series and then truncate. The width of the window determines the width of the transition band between the pass-band and stopband.

If using the window once produces some smoothness in the transfer function, then applying it twice should produce a transfer function that has a continuous first derivative and hence more convergence in the resulting Fourier series. Applying the rectangular window twice causes the sigma factors to occur twice; that is, we get the square of the sigma factors, since each application of the linear operator produces a set of $\sigma(N, k)$. Of course, the double application of the window results in a wider transition band. (See the third curves in Figures 6.2-1 and 6.2-2.) Again we can look at the process several ways. For these linear operators we may, for instance, first ask about the result of convolving the window with itself and then applying the result once. A few moments' thought shows that the convolution of the rectangular window with itself will result in a triangular-shaped window whose base is twice the original window width and whose apex is in the middle (Figure 6.3-2). Thus a triangular window is equivalent to double application of the rectangular window.

The "carpentry" possibilities of windows are endless. We can start with other than rectangular-shaped windows. Such windows are called *tinted*; a variable amount of the original function gets through at various points. One example is the preceding triangular window. In short, we can use any reasonable function that we please, supposing, for convenience, that it has unit area under the curve, and any combination, via convolutions, can be tried. The fact that we will always obtain numerical factors that multiply each frequency is derived from our earlier observation that any linear operation on a frequency gives the same frequency but changes the coefficients of the Fourier expansion by a corresponding multiplicative factor (depending on the frequency). For symmetric windows, we can expect that operating on a cosine expansion will produce only cosine terms; but if there are asym-

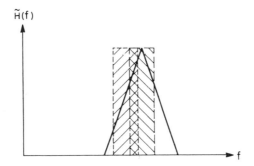

$\tilde{H}(f)$

FIGURE 6.3-2 A TRIANGULAR WINDOW FROM TWO RECTANGULAR WINDOWS

metries in the window, then the sine and cosine terms must both be considered.

Exercises

6.3-1 Smooth the rectangular pulse function with the rectangular window and then compute the Fourier coefficients directly.

6.3-2 Smooth the rectangular pulse function with the triangular window and then compute the Fourier coefficients.

6.3-3 Carry out the derivation of the window of width Δ.

6.3-4 Calculate the shape of a triple application of the rectangular window.

6.4 A DIFFERENTIATION FILTER

According to formula (6.1-3), every nonrecursive filter is the sum of a smoothing filter and a differentiator. We need, therefore, to look at the numerically difficult topic of differentiators.

Let us examine how this general method of fitting a given transfer function by a Fourier series handles the problem of designing a filter to estimate the derivative of some data. It is immediately clear from the expression for the derivative

$$\frac{d}{dt}[e^{i\omega t}] = i\omega e^{i\omega t}$$

that we must approximate the function

$$H(\omega) = i\omega$$

If we pick the coefficients of the filter to have odd symmetry, that is,

$$c_{-k} = -c_k \qquad \text{(for all } k)$$

then we see that since

$$c_k e^{ik\omega} + c_{-k} e^{-ik\omega} = c_k(e^{ik\omega} - e^{-ik\omega}) = 2ic_k \sin k\omega$$

we will have a sine series and that it will be purely imaginary as required. Thus the digital filter

$$y_n = \sum_{k=-N}^{N} c_k u_{n-k}$$

with $c_{-k} = -c_k$ leads to the sine series

$$H(\omega) = [2c_1 \sin \omega + 2c_2 \sin 2\omega + \cdots + 2c_N \sin N\omega]i$$

Examining such a filter in detail, we see that it is a linear combination of differences of symmetrically placed values of the function (estimates of the derivative)

$$c_k(u_{n+k} - u_{n-k})$$

This is what we would expect once we think about it.

It is clear that the process of differentiation amplifies high frequencies much more than low frequencies. That is why it is numerically difficult to differentiate numerical data. Since high frequency is often the noise, this means that the filter we design should probably cut off the frequencies past some value ω_c. Thus we will make the Fourier sine series approximate the function

$$H(\omega) = \begin{cases} i\omega, & |\omega| < \omega_c \\ 0, & |\omega| > \omega_c \end{cases}$$

We compute the coefficients by the usual formulas using integration by parts ($a_k = 0$ for all k):

$$b_k = \frac{1}{\pi} \int_{-\pi}^{\pi} H(\omega) \sin k\omega \, d\omega = \frac{2}{\pi} \int_{0}^{\omega_c} i\omega \sin k\omega \, d\omega$$

$$b_k = \frac{2i}{\pi} \left(\frac{\sin k\omega_c}{k^2} - \frac{\omega_c \cos k\omega_c}{k} \right)$$

As a check on the algebra, put $\omega_c = \pi$. We get

$$h_k = \frac{-2i \cos \pi k}{k} = \frac{2i}{k}(-1)^{k+1}$$

which agrees with the result of Section 4.3.

We now have the infinitely long filter that we must truncate [(b) in Figure 6.1-4], and we face the Gibbs effect. For the moment we simply use the rectangular window. Doing this produces the corresponding sigma factors in the expansion. To illustrate, we choose $f_c = \frac{2}{10}$. For values of $N = 5, 10$, we have the corresponding curves plotted in Figures 6.4-1 and 6.4-2. The straight line is the ideal, the most wiggly curve is the truncated Fourier series, which has large errors near $f = 0$, and the lower curve is the final, smoothed filter.

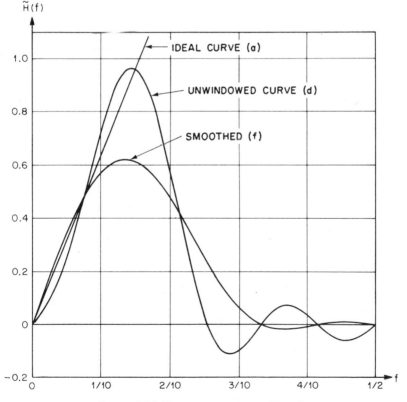

FIGURE 6.4-1 DIFFERENTIATOR WITH $N = 5$

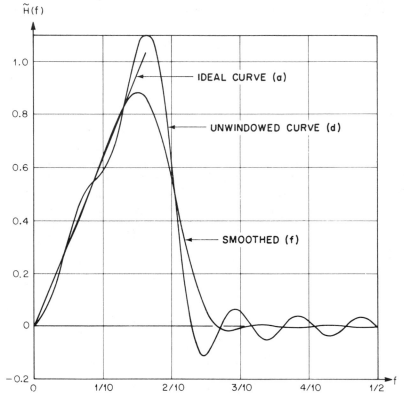

$\tilde{H}(f)$

IDEAL CURVE (a)

UNWINDOWED CURVE (d)

SMOOTHED (f)

FIGURE 6.4-2 DIFFERENTIATOR WITH $N = 10$

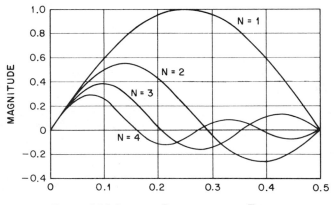

MAGNITUDE

FIGURE 6.4-3 LANCZOS DIFFERENTIATING FILTERS

Exercises

6.4-1   In [L, p. 321] a "low-noise" differentiation filter is proposed that (1) has unit slope at $f = 0$ and (2) minimizes the sum of the squares of the coefficients. Show that this filter is

$$3 \sum_{k=-N}^{k=N} k \frac{u_{n-k}}{N(N + 1)(2N + 1)}$$

(*Hint:* Use the Lagrange multiplier method.) See Figure 6.4-3.

6.4-2   Super Lanczos low-noise differentiators. In Exercise 6.4-1, if we require that the tangency at $f = 0$ be of third order, show that the first few filters have transfer functions

$$\frac{8 \sin \omega - \sin 2\omega}{6}$$

$$\frac{58 \sin \omega + 67 \sin 2 \omega - 22 \sin 3\omega}{126}$$

$$\frac{126 \sin \omega + 193 \sin 2\omega + 142 \sin 3\omega - 86 \sin 4\omega}{594}$$

See Figure 6.4-4.

6.4-3   Design a differentiator with cutoff at $f_0 = \frac{1}{10}$.

6.4-4   Design a differentiator with cutoff at f_0 arbitrary $(0 < f_0 < \frac{1}{2})$.

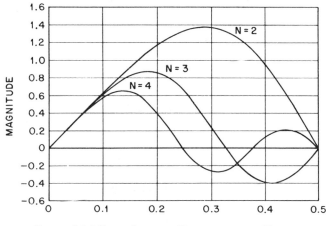

FIGURE 6.4-4 SUPER LANCZOS DIFFERENTIATING FILTERS

6.5 TESTING THE DIFFERENTIATING FILTER ON DATA

As a test of the differentiating filter, let us make up some artificial data. We first take the straight line

$$u_n = \frac{n}{50} \qquad (n = 0, 1, \ldots, 50)$$

whose derivative is, of course, $\frac{1}{50}$. To calculate the response of the filter to this input, we need only note that, since $c_k = -c_{-k}$ with $c_0 = 0$, the sum of the coefficients is zero. Therefore, we can add any constant that we like to the data without affecting the answer. Consequently we may take the straight line $u_k = k$ and test the filter at the origin of the data because it is the same as at any other place. We will then have for the output

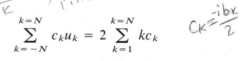

$$\sum_{k=-N}^{k=N} c_k u_k = 2 \sum_{k=1}^{k=N} k c_k$$

Without the sigma factors, we get for the value of this sum, using $N = 5$, the number $0.5165/50$ and, for $N = 10$, the number $0.5568/50$ (where we have kept the factor 50 so that the slope of the line is taken care of and the ideal value of the numerator is 1). Our choice of $N = 5, 10$ is guided by the way that we can reasonably expect the harmonics of the Fourier series to interact with the edge of the cutoff frequency at $\frac{2}{10}$.

The failure of the filter to give the value $\frac{1}{50}$ is directly traceable to the fact that at $f = 0$ the slope of the transfer function is not even near 1 (Figures 6.4-1 and 6.4-2) (highest curve), for if it were, then the calculated sums would be equal to 1.

When we use the Lanczos sigma factors (lower curve), we obtain much better results, as the figures show. The initial slope is $1.047/50$ for $N = 5$ and $1.020/50$ for $N = 10$. The straight lines show the ideal slope of $\frac{1}{50}$.

Remember, however, that we designed the filter to cope with noise. Consequently, we consider further experiments. We add random noise in two forms; first,

$$\frac{N}{50} \pm \frac{1}{100} \begin{cases} + & \text{if random number} > \dfrac{1}{2} \\ \\ - & \text{otherwise} \end{cases}$$

and a second noise

$$\frac{N}{50} + \frac{1}{100} \text{ (random number with range } -1 < u < 1)$$

The first noise may be called "shot" or "bang-bang" noise since you are always at one extreme or the other. The second noise is still very severe since noise usuallly tapers off toward the ends of its range. Figures 6.5-1 to 6.5-4 show the results of these experiments, where we have plotted (1) the original input data sloping upward, (2) ten times the derivative so that the fluctuations can be seen easily, and (3) when using the sigma factors, we also added $\frac{1}{2}$ to shift the display so that it is easily seen.

The set of experiments shows that when a filter is asked to do several things, such as both differentiate and reject a significant amount of noise. it compromises and does neither too well. The experiments also show that even simple filters can be made to do complex jobs. The behavior on a straight line was not one of the design criteria; rather it was designed to work on pure frequencies and sums of frequencies, and on them it does as

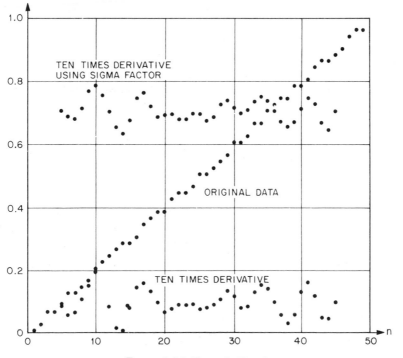

FIGURE 6.5-1 NOISE 1, $N = 5$

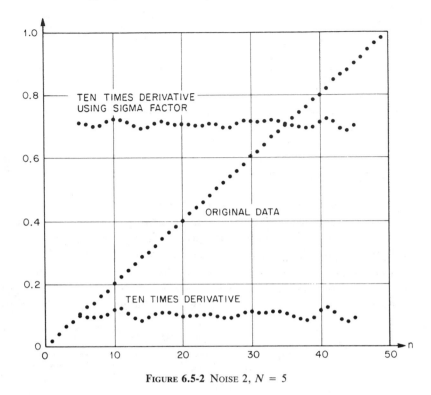

FIGURE 6.5-2 NOISE 2, $N = 5$

expected, for we can see from the transfer function exactly what happens to the amplitude of each frequency. Again, remember we have plotted 10 times the results so that the ripples can be seen.

6.6 NEW FILTERS FROM OLD ONES: SHARPENING A FILTER

It soon occurs to most students in filter theory that if using a low-pass (or high-pass or even a bandpass) filter is good, then perhaps running the data through the filter twice might be even better. As we shall soon see, this process will accomplish the following:

1. Approximately double the errors in the passband.

2. Square the error in the stopband.

3. Leave the transition band widths the same.

4. Approximately double the length of the equivalent filter (and hence the loss of data at each end).

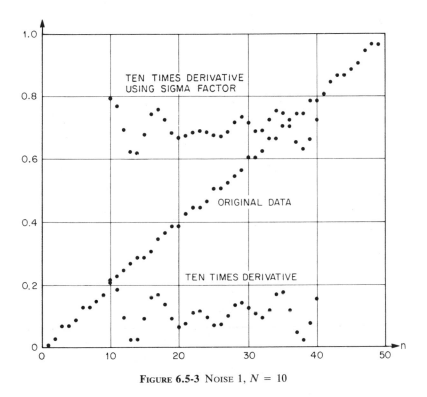

FIGURE 6.5-3 NOISE 1, $N = 10$

Of course, the equivalent single filter can be obtained by convolving the sequence of filter coefficients with itself.

Consider filtering a sequence u_n to obtain a sequence y_n. Denote the operation by

$$y_n = \tilde{H}u_n$$

Suppose that I (the identity) denotes the operation

$$u_n = Iu_n$$

Then the transfer function for I is 1, where, as usual, we use $\tilde{H}(f)$.

J. W. Tukey [T, Chapter 16] proposed (1) to run the data through a filter, (2) take the residuals

$$\epsilon_n = u_n - y_n = (I - \tilde{H})u_n$$

[where $\tilde{H}$ is the operation of applying the filter with transfer function $\tilde{H}(f)$], (3) add these to the original signal, and then (4) run the sum through the

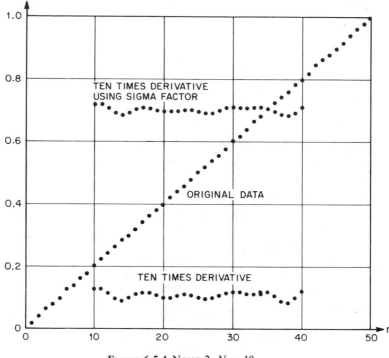

FIGURE 6.5-4 NOISE 2, $N = 10$

filter again, giving the operation

$$\tilde{H}[I + (I - \tilde{H})] = \tilde{H}(2I - \tilde{H})$$

This operator equation means that each output value of the filter $\tilde{H}$ is subtracted from $2u_n$ and the difference is then processed again by the filter $\tilde{H}$. This process Tukey calls "twicing." He also proposes further elaborations of the idea, but they will not be discussed here.

Let us examine the process more closely. For notation, we will label the original filter $\tilde{H}_{in}(f)$ and the resultant filter, after all the processing, $\tilde{H}_{out}(f)$. We now plot the *amplitude change* function $\tilde{H}_{out}(f)$ versus $\tilde{H}_{in}(f)$ (see Figure 6.6-1). We see that when $\tilde{H}_{in}(f)$ has values near 1, $\tilde{H}_{out}(f)$ has values very much closer to 1; but when $\tilde{H}_{in}(f)$ is near zero, $\tilde{H}_{out}(f)$ is approximately twice as far from zero. Thus the errors in the passband (bands) are approximately squared, whereas those in the stopband (bands) are almost doubled.

On the other hand, the opposite is true for the double use of the same filter

$$\tilde{H}_{out}(f) = \tilde{H}_{in}^2(f)$$

It is the stop part that is squared and the pass that is doubled (Figure 6.6-1).

These two special cases immediately suggest the following approach, which will (approximately) square *small* errors in both the stopband and passband of the filter. We want a curve of the amplitude change to be tangent horizontally at both 0 and 1. The polynomial having this property will be a cubic of the form

$$P(X) = p_0 + p_1X + p_2X^2 + p_3X^3$$

Applying the two conditions on this function at each end and doing some algebra, we find that we have

$$P(X) = 3X^2 - 2X^3 = X^2(3 - 2X)$$

Hence we want to use

$$\tilde{H}_{out}(f) = \tilde{H}_{in}^2(f)[3I - 2\tilde{H}_{in}(f)]$$

Three passes through the same filter result in a great sharpening up of the

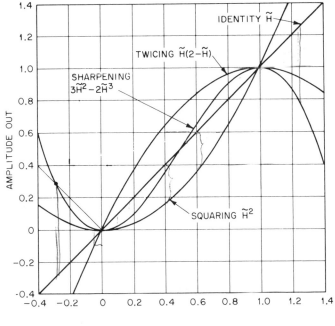

AMPLITUDE IN

FIGURE 6.6-1 AMPLITUDE TRANSFORMATION FUNCTIONS

filter. The small deviations from 0 and 1 are both squared, and the transition band (bands) width is left the same. This situation applies to low-pass, high-pass, and bandpass filters of any complexity.

So if we have a program that does some filtering job fairly well and convert it to a subroutine, then a single, short program to (1) process the signal once, (2) double this output (3) subtract each value from $3u_n$, and (4) finally pass this difference through the filter twice more will give a greatly sharpened filter of approximately three times the effective length. Of course, by suitably convolving the coefficients of the original filter, we could construct the equivalent filter and then process the signal only once. Figure 6.6-1 gives the amplitude change function. Notice that for negative values of $\tilde{H}_{in}$ and for values greater than 1, the $\tilde{H}_{out}$ produces values such that $0 \le \tilde{H}_{out} \le 1$. Thus the method squares fairly good filters. But also notice that poor filters may be made worse, since for $\tilde{H}_{in}$ either $-0.281 \ldots$ or 1.281 $\ldots$ gives $\tilde{H}_{out}$ the same magnitude of error, and beyond this range the values out are worse than the values in.

A study of Figure 6.6-2 shows that a poor filter (smoothing by $3s$, Section 3.2) is made worse in some places by sharpening. Figure 6.6-3 shows a slight improvement for smoothing by $5s$. Moreover, Figure 6.6-4 shows that smoothing by *both* $3s$ and $5s$ is a good filter and that "sharpening" it greatly improves the filter. The smoothing by $3s$ and $5s$ is simply the convolution

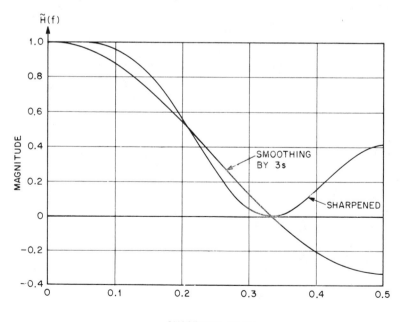

SMOOTHING BY 3s

FIGURE 6.6-2 SMOOTHING BY 3S SHARPENED

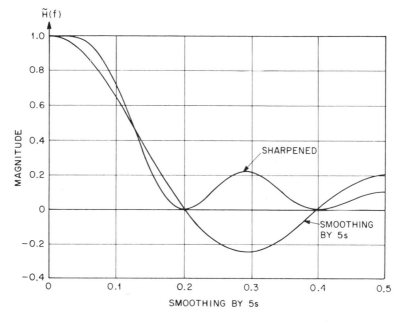

FIGURE 6.6-3 SMOOTHING BY 5s SHARPENED

of three consecutive 1s with five consecutive 1s (all divided by 3 × 5), which is the filter

$$\tfrac{1}{15}[1, 2, 3, 3, 3, 2, 1]$$

Chapter 9 will show that if we tried to design a filter of the same quality as the combination, we would find (for reasonably good filters) that we could design one of approximately double the length of the original filter. The procedure, of course, requires complete redesigning of the filter, as well as extra programming. Again, if the filter were on an integrated-circuit chip, then the proper use of three such chips would probably be cheaper than the redesign and construction of the new chip. In any case, the concept of the amplitude transformation process and its corresponding picture sheds much light on the general area of the combination of copies of the same or even different filters.

Exercises

6.6-1 Using the methods of this section, show that if we require second-order tangency at the two ends of the interval, then the corresponding amplitude change function is

$$H^3[10 - 15H + 6H^2]$$

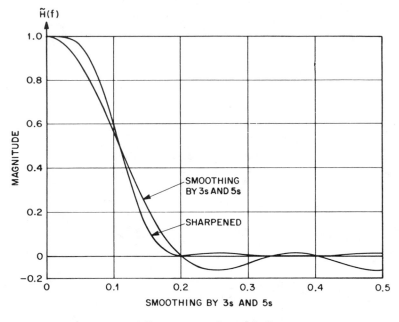

FIGURE 6.6-4 SMOOTHING BY 3S AND 5S SHARPENED

The corresponding points of no improvement are $-0.264.\ .\ .$and $1.264.\ .\ .\ .$

6.6-2   Show that when we require nth-order tangency at the two ends, the amplitude change function has the form

$$H^{n+1} \sum_{k=0}^{k=n} \frac{(n + k)!}{k!n!} (1 - H)^k$$

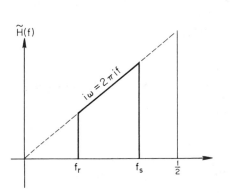

FIGURE 6.7-1 BANDPASS DIFFERENTIATOR

6.7 BANDPASS DIFFERENTIATORS

It is hardly more than an exercise to design a nonrecursive bandpass differentiator (Figure 6.7-1). We begin with equation (4.9-8); since $g(f) = 2\pi i f$.

$$
\begin{aligned}
c_k &= \int_{-1/2}^{1/2} g(f) e^{-2\pi i f k} \, df \\
&= \int_{-1/2}^{1/2} 2\pi i f [\cos 2\pi kf - i \sin 2\pi kf] \, df \\
&= 4\pi \int_0^{1/2} f \sin 2\pi kf \, df \\
&= 4\pi \left\{ f \left. \frac{-\cos 2\pi kf}{2\pi k} \right|_{f_p}^{f_s} + \frac{1}{2\pi k} \int_{f_p}^{f_s} \cos 2\pi kf \, df \right\} \\
&= \frac{2}{k} \{f_p \cos 2\pi kf_p - f_s \cos 2\pi kf_s\} + \left\{ \frac{\sin 2\pi kf_s - \sin 2\pi kf_p}{\pi k^2} \right\}
\end{aligned}
$$

$$(6.7\text{-}1)$$

Remember from Section 4.9 that the coefficients of the digital filter have the opposite signs to these Fourier coefficients in this case of a differentiator.

To check the result (partially), we note that if $f_p = 0$, then we should have the result of Section 6.4 (which in its turn was checked against an earlier computation). To do this we have to convert to radians and remember from (4.9-1) that for $k > 0$

$$
c_k = \frac{-i b_k}{2}
$$

6.8 MIDPOINT FORMULAS

Formulas whose outputs fall midway between the input data are widely used in numerical work. We analyzed the transfer function for several such formulas and found that they involved frequencies of half the odd integers (see Section 3.7). Experts in digital filter theory tend to object to such formulas because the output occurs at places where the input did not occur, and they ignore the simple fact that numerical analysts find such formulas very useful and are able to handle the midpoints quite easily.

The midpoint formulas are mainly used in problems involving the use of derivatives since they so clearly have great accuracy in such situations compared with the estimates at the sample points.

In Section 4.4 we laid the elements for the design of midpoint filters. There we showed that if we extend the definition of an odd function from the interval $0 \leq \omega \leq \pi/2$ to the interval $0 \leq \omega \leq \pi$ using symmetry about $\pi/2$, we will get only odd harmonics of the sines. Similarly, for even functions, using odd symmetry about $\pi/2$ will produce odd harmonics in the cosines.

To adapt these formulas to our needs, we simply write

$$(2k + 1)\omega = (k + \tfrac{1}{2})(2\omega) = (k + \tfrac{1}{2})\omega'$$

and then set $\omega' = 2\omega$. The unit spacing of the samples now represents $\tfrac{1}{2}$ the spacing in the original ω. Thus the folding frequency is now at the old $f = \tfrac{1}{4}$. In the case of the sines the extended function now falls exactly on top of the original piece of the function, and for the cosines it is the same size but of opposite sign for each value of f. We know how to get the expansions in the half-odd integers, and thus the midpoint formulas we want.

Let us carry out some of the initial details in the case of a differentiator (Figure 6.8-1). First, we have

$$g(\tfrac{1}{2} - f) = g(f) \qquad (\,|f| \leq \tfrac{1}{4})$$

$$g(-f) = -g(f)$$

$$c_k = \int_{-1/2}^{1/2} g(f)e^{-2\pi i k f}\,df$$

$$= \int_{-1/2}^{1/2} g(f)[\cos 2\pi k f - i \sin 2\pi k f]\,df$$

$$= -2i \int_{0}^{1/2} g(f) \sin 2\pi k f\,df$$

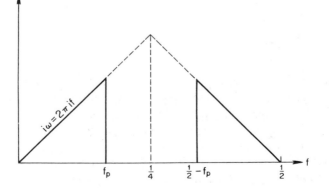

FIGURE 6.8-1 INITIAL FUNCTION

For odd harmonics

$$c_{2k+1} = -4i \int_0^{1/4} g(f) \sin 2\pi(2k+1)f \, df$$

Now set $2f = f'$, drop the primes, and redefine $g(f)$:

$$c_{2k+1} = -2i \int_0^{1/4} g(f) \sin 2\pi(k + \tfrac{1}{2})f \, df$$

Next use $g(f) = i\omega = 2\pi i f, \ 0 \le f \le f_c$:

$$c_{2k+1} = 4\pi \int_0^{f_c} f \sin 2\pi(k + \tfrac{1}{2})f \, df$$

and we have our coefficients *when* we remember that the filter coefficients in this case have the opposite sign to those of the expansion (Section 4.9).

An interesting case arises when $f_c = \tfrac{1}{2}$. The original function being expanded looks like the sawtooth of Figure 4.4-1 and has no discontinuities; hence the convergence, by Section 4.6, is like $1/k^2$.

7

Smooth Nonrecursive Filters

7.1 OBJECTIONS TO RIPPLES IN A TRANSFER FUNCTION

The filters designed in the previous chapter had ripples in the transfer function, and, for some purposes, this situation is undesirable. For instance, ripples in the passband are objectionable in cases in which the filtering is cascaded, one filter after another. If the signal passes through M identical filters, then any peak of $1 + \delta$ becomes

$$(1 + \delta)^M$$

which can cause overload (overflow in a fixed-point computer) for sufficiently large M. In sophisticated signal processing the cascading of filters is not unusual (for example, in long-distance telephone communication).

Again, if the problem is one in which the signal (information) is masked by a large amount of noise, then after filtering out the noise, any small peaks that are left in the spectrum of the signal might be either from the original signal or from ripples in the transfer function used in the filtering process. Careful analysis could separate the two, but the problem can also be avoided if a class of filters that vary smoothly rather than ripple is used. By "vary smoothly" we mean that the filters are *monotone* over long intervals of the frequency band.

An example of a monotone filter occurs when we interpolate midpoints between some equally spaced data. Such problems arise in many places. For instance, in demography data is sometimes published as applying for the year end and sometimes for the midyear; to combine data from two such different sources, it is necessary to interpolate midpoints in one of the sets of data, and one would like, typically, a monotone interpolation value filter. Again, as noted in Section 6.8, if we use the derivative at the half points we may later need to interpolate data at these half points.

Assuming the usual polynomial approximation, we get for linear interpolation for the midpoint

$$y_{n+1/2} = \tfrac{1}{2}[u_{n+1} + u_n]$$

If we normalize our problem and imagine we are interpolating at the origin, then we have the two-term formula

$$y_0 = \tfrac{1}{2}[u_{1/2} + u_{-1/2}]$$

which we write in the symbolic form

$$(\tfrac{1}{2})[1, \ 1]$$

Passing third-, fifth-, and seventh-degree polynomials through four, six, and eight terms, respectively, gives the corresponding symbolic formulas:

$$(\tfrac{1}{16})[-1, \ 9, \ 9, \ -1]$$

$$(\tfrac{1}{256})[3, \ -25, \ 150, \ 150, \ -25, \ 3]$$

$$(\tfrac{1}{2048})[-5, \ 49, \ -245, \ 1225, \ 1225, \ -245, \ 49, \ -5]$$

Their corresponding transfer functions are shown in Figure 7.1-1. We see that, as expected (Section 3.3), the higher-order polynomial has the higher-order tangency at the origin and comes closer to being an all-pass filter.

If we design a filter that removes more of the high frequency, we will find that making it tangent at the folding frequency $f = \tfrac{1}{2}$ as well as at the origin will, for low-order filters, give all positive coefficients. This situation reminds us of the following easily proved theorem:

THEOREM. *Having all the coefficients positive is the necessary and sufficient condition for a smoothing filter also to have the property that if one function is always larger than a second, then its output will be larger as well.*

But for such a filter this statement means that, if there is a local peak, the interpolated value must be less than the maximum of the values (it is a weighted average with positive weights); thus peaks are cut off (and valleys

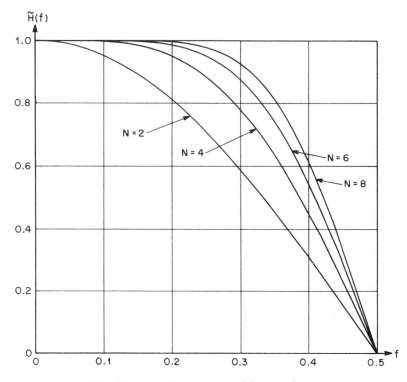

FIGURE 7.1-1 TRANSFER FUNCTIONS FOR MIDPOINT INTERPOLATION

are filled in). The peaks and valleys can be reasonably accurately portrayed by the filter only if we allow high frequencies to pass and thereby allow positive and negative coefficients in the filter. The alternating signs in the formulas we have derived mean that the sum of the squares of the coefficients will be large, and hence a large propagated roundoff will occur (Section 1.7).

This is one of the many difficulties of filtering: we simply cannot have everything that we want. By removing the high-frequency noise, we tend to force the response to changes in the signal to be less; we tend to flatten out peaks and fill in valleys and other sudden changes. If we insist on keeping the small details of the data, we must also keep the high-frequency noise.

Exercises

7.1-1 Prove the theorem in this section.

7.1-2 Derive the formulas for polynomial midpoint interpolation.

7.2 SMOOTH FILTERS

Let us once again review where we are in the topic of filters. Given a symmetric ($c_k = c_{-k}$), nonrecursive digital filter operating on equally spaced data, u_n, from some source, we compute a y_n by the formula

$$y_n = \sum_{k=-N}^{N} c_k u_{n-k} \qquad (7.2\text{-}1)$$

The eigenfunctions of linear problems are the complex exponentials $e^{2\pi i f n}$ (Section 2.5). The use of the eigenfunctions leads to the transfer function (eigenvalues)

$$\tilde{H}(f) = c_0 + 2 \sum_{k=1}^{N} c_k \cos 2\pi k f$$

We now show the well-known fact that

$$\cos k\theta = \text{polynomial in } \cos\theta \text{ of degree } k$$

To prove it, we begin with the obvious fact that

$$e^{in\theta} = [e^{i\theta}]^n$$

Next, write this equation in the complex form

$$\cos n\theta + i \sin n\theta = [\cos\theta + i \sin\theta]^n$$

Expand the right side binomial, and take the real part of the equation

$$\cos n\theta = \sum_{k=0}^{n/2} C(n, 2k) \cos^{n-2k}\theta (i \sin\theta)^{2k}$$

where, of course, the summation is cut off when $2k$ exceeds n, because then the binomial coefficients are zero. Since $i^{2k} = (-1)^k$ and

$$\sin^{2k}\theta = [\sin^2\theta]^k = [1 - \cos^2\theta]^k$$

we have the desired polynomial in powers of $\cos\theta$. In practice we will not carry out the details this way.

Returning to our transfer function (7.2-1)

$$\tilde{H}(f) = c_0 + 2 \sum_{k=1}^{N} c_k \cos 2\pi k f$$

we can use the preceding result to get, for suitable b_k's,

$$\tilde{H}(f) = \sum_{k=0}^{N} b_k [\cos 2\pi f]^k \qquad (7.2\text{-}2)$$

This is a shift from using $\cos k\theta$ to using powers of $\cos \theta$.

We next make the transformation of the independent variable

$$\cos 2\pi f = t \qquad (7.2\text{-}3)$$

As f goes from 0 to $\frac{1}{2}$, t goes from 1 to -1, and we will have the polynomial in t, from (7.2-2)

$$\tilde{H}(f) = \sum_{k=0}^{N} b_k t^k$$

as an equivalent transfer function. However, note that the transformation is a nonlinear stretching of the frequency axis. By working in the variable t, which is $\cos 2\pi f$, we will be representing the transfer function in terms of powers of $\cos 2\pi f$, rather than in cosines of multiple angles as we have been doing up to now. The original low-pass filter, because of the reversal of the axis in the transformation, now "looks like" a high-pass filter in the t variable. The transformation uses only $0 \le f \le \frac{1}{2}$. But by symmetry if we handle this part we will, when we transfer back to the variable f, be able to reconstruct the entire transfer function.

To get started on the design, we choose the function

$$g(t) = [1 + t]^p [1 - t]^q \qquad (7.2\text{-}4)$$

with p and q as parameters (Figure 7.2-1). Clearly, this function has a zero of order p at $t = -1$ and a zero of order q at $t = 1$. Integrating this function of t gives an arbitrary constant of integration, which we will fix so that at $t = -1$ the integrated function is zero. Next, we will calculate the value of the function at $t = 1$ and divide by this number (normalize) so that the final function has the value 1 at $t = 1$ (Figure 7.2-1). This, then, is our transfer

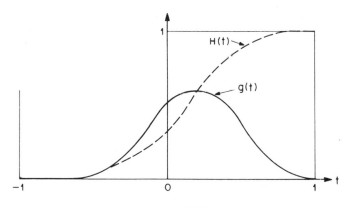

FIGURE 7.2-1

function represented in the variable t.

$$H(t) = \frac{\int_{-1}^{t} (1 + t)^p (1 - t)^q \, dt}{\int_{-1}^{1} (1 + t)^p (1 - t)^q \, dt} \tag{7.2-5}$$

Those familiar with the function may recognize it as the incomplete Beta function in disguise; the normalizing factor is the corresponding complete Beta function.

The denominator can be reduced to the standard Beta function by the transformation

$$1 + t = 2u$$

$$\int_{-1}^{1} (1 + t)^p (1 - t)^q \, dt = \int_{0}^{1} (2u)^p 2^q (1 - u)^q 2 \, du$$

$$= 2^{p+q+1} \int_{0}^{1} u^p (1 - u)^q \, du$$

$$= 2^{p+q+1} \frac{p! q!}{(p + q + 1)!}$$

The last step is the standard Beta function and can be evaluated by integration by parts.

After integration, we have a polynomial in t (our transfer function) that has a $(p + 1)$th-order zero at $t = -1$ and the value 1 at $t = 1$, along with q derivatives that are zero at $t = 1$. The reason that the order of the highest vanishing derivative rises by 1 is that the process of integration increases the tangency at the origin and at the terminal value.

This function in t has the coefficients b_k in the transfer function (no relation to the Fourier coefficients b_k), and the simple substitution of $t = \cos 2\pi f$ will bring us back to the frequency variable. To follow this transformation, we have

$$\tilde{H}(f) = H(\cos 2\pi f) = \frac{\displaystyle\int_{-1}^{\cos 2\pi f} (1 + t)^p (1 - t)^q \, dt}{\displaystyle\int_{-1}^{1} (1 + t)^p (1 - t)^q \, dt}$$

We need only examine the behavior of the numerator. We make the transformation in the integral back to f (7.2-3)

$$t = \cos 2\pi f$$

Therefore, using the half-angle formulas (and $\sin 2\pi f = 2 \sin \pi f \cos \pi f$),

$$\int_{-1}^{\cos 2\pi f} = \int_{1/2}^{f} (1 + \cos 2\pi f)^p (1 - \cos 2\pi f)^q (-2\pi \sin 2\pi f) \, df$$

$$= 2^{p+q+2} \pi \int_{f}^{1/2} (\cos \pi f)^{2p+1} (\sin \pi f)^{2q+1} \, df$$

$$= 2^{p+q+2} \pi \int_{f}^{1/2} \left[\sin \pi \left(\frac{1}{2} - f \right) \right]^{2p+1} [\sin \pi f]^{2q+1} \, df$$

Since $\sin \pi f$ is approximately πf near a zero of a sine function we see the doubling in the order of tangency at both ends of the interval due to the "compression" caused by the transformation back to f (7.2-3).

A simple method, which will be developed in the next section, enables us to make the transformation from the powers of $\cos 2\pi f$ back to the Fourier series representation of the transfer function to get the c_k coefficients of the digital filter that we started out to find.

Returning to the design problem, we equate $g'(t)$ to zero and find that the maximum of $g(t)$ and hence the inflection point of the function $H(t)$ in the t domain occurs at

$$\frac{p - q}{p + q}$$

Increasing both parameters p and q proportionately gives a narrower transition zone in $H(t)$.

The original curve $g(t)$ is a polynomial in t with all its features assigned at the ends of the interval; thus there are no "wiggles" in both the $g(t)$ and $H(t)$ curves between the endpoints. The transformation back to the f variable is monotone and only stretches the independent variable axis, which means that it cannot put in maxima and minima. The flatness of the cosine curve of the transformation doubles the tangencies at the ends.

We have yet to do the transformation back to the Fourier series notation, but this step is only a change of notation and does not affect the shape of the curve. Thus we obtain a smooth transfer function. The direct design method using the Fourier series terms of the form $\cos 2\pi kf$ tends to produce wiggles in the transfer function. It is to escape these wiggles that we have made the fit very good at the ends of the interval and let the rest fall where it may, using only the degrees of the tangencies at the ends to control where the filter cutoff occurs.

Exercises

7.2-1 Derive the formula for the location of the inflection point.

7.2-2 Convert $1 + t + t^2 + t^3$ to a Fourier series, where $t = \cos 2\pi f$.

7.2-3 Show that $\beta(m, n) = \displaystyle\int_0^1 x^m (1 - x)^n \, dx = \dfrac{m!\,n!}{(m + n + 1)!}$

7.3 TRANSFORMING TO THE FOURIER SERIES

This section discusses the valuable method of how to get from a power series in $\cos \theta$ to a Fourier series in $\cos k\theta$. Furthermore, we want to do it easily on a computing machine. Thus we look for a recursive way of doing the conversion that will make the programming easy.

Let the power series in $\cos \theta$ (7.2-1)

$$\sum_{k=0}^{N} b_k [\cos \theta]^k \qquad (7.3\text{-}1)$$

be written in the usual "chain" form.

$$b_0 + \{\cdots + \cos \theta[b_{N-2} + \cos \theta(b_{N-1} + b_N \cos \theta)] \cdots\}$$

Starting at the inner parentheses with the last two terms

$$b_{N-1} + b_N \cos \theta \qquad (7.3\text{-}2)$$

we multiply by $\cos \theta$ and add the next lower coefficient. Then we again multiply by $\cos \theta$ and add the next lower coefficient, and so on.

The starting two terms (7.3-2) are in the form of a Fourier series and thereby form the basis for an induction (a recursive process). Consequently for the induction process, we assume that at each stage we have a Fourier series with the coefficients given and show that the next stage is also a Fourier series. To show this, we multiply the current Fourier series in the induction process by $\cos \theta$ and use the formula for the product of two cosines:

$$\cos \theta \cos n\theta = \tfrac{1}{2}[\cos(n + 1)\theta + \cos(n - 1)\theta]$$

From any one coefficient with index n greater than zero, we get two terms; the first is one unit in frequency higher and the second is one frequency lower, each with the coefficient $\tfrac{1}{2}$. The constant term of index $n = 0$ becomes a cosine term with the full coefficient. Finally, the next lower coefficient b_k is added to the constant term that came from the earlier calculations of the term with index equal to 0. See Figure 7.3-1, where the dots mean the corresponding coefficients of the Fourier terms of increasing frequency toward the right and the small figures alongside the arrows indicate the multipliers. If this simple process is repeated $N - 1$ times, we end up with the Fourier series that is equivalent to the original power series in $\cos \theta$. The coefficients in the corresponding complex Fourier series are the coefficients c_k of the filter.

Notice that the process has a number of nice properties. First, it is robust; the division by 2 at the various stages produces no roundoff in *floating-point binary* arithmetic and tends to decrease errors. Second, it is simple to program, and involves no tables and little arithmetic beyond additions. Finally, this same process will be used in another situation; thus it should

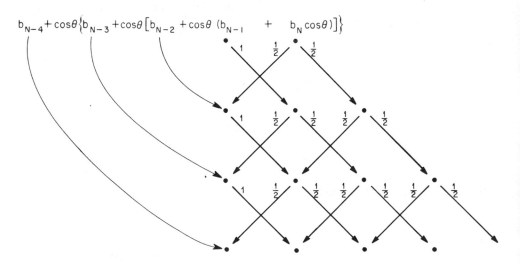

FIGURE 7.3-1 ONE STAGE IN TRANSFORMING TO THE FOURIER SERIES

be understood. Evidently, it depends on the simple trigonometric formula for the product of two cosines and nothing more beyond the organization of the computation into a regular form.

Exercises

7.3-1   Convert $\cos^5 \theta$ to an equivalent Fourier series.

7.3-2   Convert $\cos^4 \theta + \cos^2 \theta + \cos \theta + 1$ to an equivalent Fourier series.

7.3-3   Convert $\cos^5 \theta - 5 \cos^3 \theta + 5 \cos \theta$.

7.4 POLYNOMIAL PROCESSING IN GENERAL

While on the general topic, which might be called "the engineering of algebra," we will pause to organize the handling of polynomials that our design process requires. Again, we will see that recursive methods are preferable to methods that might come to mind as a result of taking the usual mathematics courses. Generally, such courses concentrate on ideas and ignore the operations of carrying them out. It is clearly necessary to understand what we are doing before mastering the way we do it; so we are not faulting mathematics courses by saying that they neglect the engineering of algebra. Yet to do the extensive symbol manipulation that often arises in science and engineering, we cannot overlook the practical situation. We shall organize matters both for the computer and for small designs done by hand.

How, for example, do we do some of the algebra required in the preceding design? We began with the expression (7.2-4)

$$[1 + t]^p [1 - t]^q \qquad (7.4\text{-}1)$$

and obtained a polynomial in t. Let us suppose that $p \le q$. It is just as easy to handle the opposite case. We write

$$q = p + k \qquad (k \ge 0)$$

Therefore, we have as the starting point

$$[1 + t]^p [1 - t]^p [1 - t]^k = [1 - t^2]^p [1 - t]^k \qquad (7.4\text{-}2)$$

The first bracket on the right side can be expanded by the standard binomial process to get the series in powers of t^2. Notice that: (1) viewed as a polynomial in t, every other term is zero; (2) by the usual process, the binomial coefficients are recursively calculated; and (3) a zero coefficient indicates the end of the recursion. The maximum binomial coefficient is the middle one(s), but it refers to index p, whereas we have an expansion in t up to powers $2p$.

To multiply this expansion by $(1 - t)$, we need only copy the set of coefficients shifted one place to the right and then subtract from the original coefficients. If $k > 0$ the first step merely combines nonzero numbers with the zeros and actually does no arithmetic, but fits into the recursion. This process of shifting the set of coefficients one to the right and then subtracting should be done exactly k times to allow for the factor $(1 - t)^k$. It can be done "in place" in a computer without wasting a lot of storage.

To integrate the resulting polynomial, we merely divide the kth coefficient by $k + 1$ (and, of course, increase in our minds the power of t that it represents). Evaluating this polynominal at $t = -1$ consists of simply adding the coefficients with alternating sign and determining the constant of integration term C that is needed to make the integral (7.2-5) zero at $t = -1$.

Up to now we can use any (nonzero) multiple of the polynomial that we please, since the final normalization (the denominator of 7.2-5) to make the polynomial have the value 1 at $t = 1$ will remove this multiplying factor.

Now that we have the constant term we can evaluate the polynomial at $t = 1$. To do so, we merely sum all the coefficients. Finally, dividing all the coefficients by this number normalizes the integral so that it has the value 1 at $t = 1$. Thus we have our transfer function in polynomial form.

Again, we see that simple organization based on recursive methods makes larger masses of algebra practical, both on a machine and by hand.

Exercise

7.4-1   Discuss in detail the case when $p \geq q$.

7.5   THE DESIGN OF A SMOOTH FILTER

Suppose we want to design a smooth filter that passes the lower one-third of the Nyquist interval, that is, up to $\pi/3$, and stops frequencies in the upper part. Going to the t coordinate (7.2-3), the edge of the filter occurs at

$$\cos\left(\frac{\pi}{3}\right) = t = \frac{1}{2}$$

We need to select p and q such that (at least approximately) (7.2-6)

$$\frac{p - q}{p + q} = \frac{1}{2}$$

The two choices $p = 3$, $q = 1$ and $p = 6$, $q = 2$ differ only in the sharpness of the cutoff.

Case 1. $p = 3$, $q = 1$ (almost trivial)
The starting polynomial is (7.4-2)

$$(1 + t)^3(1 - t) = (1 - t^2)(1 + t)^2$$

Powers of $t \rightarrow$		1	t	t^2	t^3	t^4	t^5
$1 - t^2$	$=$	1	0	-1			
Shift right			1	0	-1		
$(1 - t^2)(1 + t)$	$=$	1	1	-1	-1		
Shift right			1	1	-1	-1	
$(1 - t^2)(1 + t)^2$	$=$	1	2	0	-2	-1	
Integrate $\rightarrow$		C	1	1	0	$-\frac{1}{2}$	$-\frac{1}{5}$
Fix constant $\rightarrow$		$\frac{1}{10}[3$	10	10	0	-5	$-2]$
Scale		$\frac{1}{16}[3$	10	10	0	-5	$-2]$

These are the b_k's in the $\cos(2\pi f)$ expansion (7.3-1). To avoid fractions in the hand conversion to the c's of the Fourier expansion, we write the polynomial in the form (multiply and divide by $2^4 = 16$)

$$(\tfrac{1}{16})^2[48 \quad 160 \quad 160 \quad 0 \quad -80 \quad -32]$$

The conversion process from the b_k's to the Fourier coefficients a_k's is according to Section 7.3:

-80	-32				
-16	-80	-16			
0					
-16	-80	-16			
-40	-8	-40	-8		
160	-16				
120	-24	-40	-8		
160	-20	-4	-20	-4	
-12	120	-12			
148	100	-16	-20	-4	
48	148	-10	-2	-10	-2
50	-8	50	-8		
98	140	40	-10	-10	-2

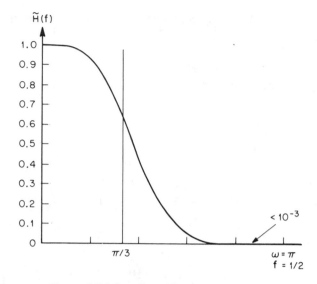

FIGURE 7.5-1 LOW-PASS MONOTONE FILTER

Thus halving all but the zero coefficient the filter is

$$(\tfrac{1}{16})^2[-1 \quad -5 \quad -5 \quad 20 \quad 70 \quad 98 \quad 70 \quad 20 \quad -5 \quad -5 \quad -1]$$

As a result, we have the monotone filter shown in Figure 7.5-1.

Exercise

7.5-1   Do the case $p = 6$, $q = 2$ and compare the transfer function plots.

7.6 SMOOTH BANDPASS FILTERS

To extend these ideas to a bandpass filter, we need only start with a different function. Again we want zeros at both ends, but this time we also want an odd-order zero where the middle of the passband occurs. Thus we start with a polynomial of the form [Figure 7.6-1(a)]

$$(t + 1)^p(t - t_0)^{2q+1}(t - 1)^r$$

where we have not as yet picked the value t_0. The choice of the exponents is clear once we realize that the p and r tend to control where the passband will be and that the $2q + 1$ controls the flatness of the passband.

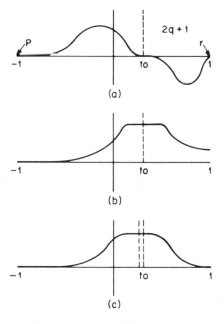

FIGURE 7.6-1

Next, we expand this polynomial first by using the binomial expansion on the middle term to get a polynomial in t_0. Clearly, a computer is needed to do all the arithmetic! Each coefficient is a polynomial in t; and using the same methods illustrated in the previous section, we arrange each coefficient of t_0 as a polynomial in t.

Integration of these polynomials from -1 to t will show that the sum of all the integrals will usually not be zero at $t = 1$ for our initial choice of t_0 [Fig. 7.6-1(b)]. Each of the coefficients of the powers of t_0 can be evaluated at $t = 1$, and we have a polynomial in t_0 with numerical coefficients. The bisection method (or Newton's method) can be used to find the *simple* zero at t_0 such that the integral will be zero at $t = 1$. After finding this t_0, we have our integrated polynomial except that we must still normalize it so that at the value t_0 it has the value 1 [Fig. 7.6-1(c)].

Once we have the t_0 and the normalization, the filter process design is much as before. Conversion to the Fourier series form and then to the ultimate digital filter consists of the same processes used earlier. It is clear that these designs are best done on a computer and that the actual programming is recursive and short. The actual machine time used is not high either.

8

The Fourier Integral and the Sampling Theorem

8.1 INTRODUCTION

The Fourier series is useful for handling both periodic functions and functions with a finite range of independent variables (since this finite range may be extended to the whole line by defining a periodic extension of the function). However, periodic functions are comparatively rare in practice; and in order to consider many of the applications of digital filters, a more general class of functions is necessary. This situation, in turn, requires the development of a more general mathematical tool to handle the nonperiodic functions, that is, the Fourier integral. A simple example of a nonperiodic function of time whose components are periodic is

$$y(t) = \cos t + \cos \sqrt{2} \, t$$

Since 1 and the $\sqrt{2}$ are not commensurate, this function cannot repeat itself exactly no matter how far we go in the variable t.

The formal definition of the Fourier integral representation for $g(t)$ is

$$g(t) = \int_{-\infty}^{\infty} G(f)e^{2\pi i f t} \, df$$

Clearly, in the Fourier integral, there are a noncountable number of frequencies, one corresponding to each real number f.

In practice, we will usually deal only with frequencies in a band, for, as we saw in Chapter 2, the process of sampling at a sequence of equally spaced points produces an effect that can be viewed as aliasing each frequency into a frequency lying in the Nyquist interval. Such functions are called *band limited* because, for practical purposes, once sampled then all their frequencies are within a band such that at least two samples in the highest frequency present is the rule. Other physical examples of continuous functions that are almost band limited are common; for instance, the typical hi-fi system has a cutoff at the low end of a few tens to possibly a hundred hertz, and at the upper end the cutoff frequency may be as high as 20,000 Hz.

One purpose of this chapter is to provide a mathematical bridge between the original continuous function that we customarily think about and the sampled function that must be dealt with when using a digital filter. We also need to know how to go from the sampled function that is band limited back to the original continuous function—how to interpolate the function between the given equally spaced samples. For these reasons, we must look at the famous *sampling theorem*.

Finally, we examine the effect of taking a finite piece of a continuous function from what is potentially an infinitely long function and various ways of using windows to reduce this effect.

8.2 SUMMARY OF RESULTS

This section summarizes the rest of the chapter and does not try to prove the results rigorously, since rigorous proofs would draw us deep into the details of mathematics and obscure what is happening. Partial proofs of some of the results will be given later in this chapter.

The first thing to show is that for any reasonable function $g(t)$ there is a representation in the form of a Fourier integral

$$g(t) = \int_{-\infty}^{\infty} G(f)e^{2\pi ift} \, df \qquad (8.2\text{-}1)$$

where we also have

$$G(f) = \int_{-\infty}^{\infty} g(t)e^{-2\pi ift} \, dt \qquad (8.2\text{-}2)$$

These formulas are sometimes called the *Fourier inversion formulas*. In analogy with the Fourier series, from which we shall derive the Fourier integral, we have the *density function* $G(f)$ corresponding loosely to the c_k of the complex form of the Fourier series (Section 4.9); the density $G(f)$ times the interval Δf gives the amount in measurable units. The function $G(f)$ is called the *transform* of $g(t)$. We shall adopt the customary usage of lowercase letter in the time domain and the corresponding capital letter in the frequency domain. As can be seen, the only difference (in the f notation but *not* in the ω notation) in the two equations is that one has a $-i$ in place of i. We need not, of course, use frequency and time—any pair of variables so related can be handled by these methods. In quantum mechanics these are called "conjugate variables."

Aliasing plays a fundamental role in sampling. The earlier discussion of aliasing with respect to Fourier series (Section 2.2) was in no way dependent on the discrete frequencies that occurred, and everything goes through exactly the same for the continuous range of frequencies used in the Fourier integral. If the frequencies are limited to a symmetric band about the origin, as is usual,

$$-F \leq f \leq F$$

of width $2F$, and the sampling is done at an interval of Δt between the samples, then we have the important relation

$$2F \, \Delta t < 1 \qquad\qquad (8.2\text{-}3)$$

as a necessary condition to avoid aliasing (the inequality rather than the equality is necessary to avoid the end frequencies that are theoretically troublesome but that are easy to cope with in practice). The maximum sampling interval $1/(2F)$ is related to the folding or Nyquist frequency. The common way of expressing this relationship is to say, as before, that we must have at least two samples in the highest frequency present in order to avoid aliasing. For our standard $\Delta t = 1$ we have $|F| < \frac{1}{2}$.

Exercises

8.1-1   Write the Fourier integral in the angular frequency notation. How should the coefficient 2π be handled?

8.1-2   If there are 100 samples per second, what is the folding frequency?

8.1-3   If the measurements are made yearly what is the folding frequency?

8.3   THE SAMPLING THEOREM

Some questions that naturally occurs are: Are the samples as good as the original function? Is it possible to reconstruct a band-limited function from its samples, *provided* we observe the sampling restriction? The fundamental result that it *is* possible is called the *sampling theorem*, which is so important that two different, partially rigorous proofs will be given in the hope that doing so will make the theorem understandable.

As background for this theorem let us look at the corresponding result in polynomial interpolation, since interpolation is really what we are doing in the sampling theorem. From the infinite number of discrete samples, we try to reconstruct the value of the band-limited function at any given point of the function. In the polynomial case for a finite number of points, t_1, t_2, ..., t_n, we use the Lagrange interpolation formula. The standard notation is (see Section 2.6)

$$\pi(t) = (t - t_1)(t - t_2) \cdots (t - t_n), \qquad \pi_k(t) = \frac{\pi(t)}{t - t_k} \qquad (8.3\text{-}1)$$

where $\pi_k(t)$ is a polynomial of degree $n - 1$ and is the product of all the differences *except* the kth one. Next consider the expression

$$\frac{\pi_k(t)}{\pi_k(t_k)} \qquad (8.3\text{-}2)$$

This expression is zero at all the sample points t_j *except* the kth one, at which it is exactly 1. Once we understand this property, it is easy to see that the function

$$g(t) = \sum_{k=1}^{n} g(k) \frac{\pi_k(t)}{\pi_k(t_k)} \qquad (8.3\text{-}3)$$

has the value $g(k)$ at the kth point and is therefore the interpolating polynomial of degree $n - 1$ that passes through the given values $g(k)$ (within roundoff).

If the frequency approach is used instead of the polynomial approach, corresponding to (8.3-2), we are led to consider the function (for unit spacing)

$$\frac{\sin \pi(k - t)}{\pi(k - t)} \qquad (8.3\text{-}4)$$

which for $t = k$ takes on the value 1 and for all other integers has the value 0. Thus the formal expression

$$g(t) = \sum_{k=-\infty}^{\infty} g(k) \frac{\sin \pi(k - t)}{\pi(k - t)} \qquad (8.3\text{-}5)$$

clearly passes through the sample values $g(k)$. We say "formal" because we have no indication at present that the infinite series will converge except at the sample points.

The original function from which we obtained the samples $g(k)$ was assumed to be band limited; is the function $g(t)$ also band limited? It is easy to show that it is by direct integration. Taking unit spacing for our samples, the particular band-limited function that is 1 in the band $-\frac{1}{2}$ to $\frac{1}{2}$ and 0 outside leads to

$$\int_{-1/2}^{1/2} e^{2\pi i f t}\, df = \frac{e^{\pi i t} - e^{-\pi i t}}{2i\pi t} = \frac{\sin \pi t}{\pi t} \qquad (8.3\text{-}6)$$

Therefore, the function on the right side of this equation is band limited and lies in the proper band. Shifting the independent variable t a fixed amount k does not change the frequencies in the function, and we conclude (from the band-limited property of the individual terms of the formal sum) that the sum itself is band limited. This is our first, rather formal proof of the sampling theorem; from the samples $g(k)$ of a band-limited function we can reconstruct the original function. Some attention to the uniqueness of the function is, of course, necessary.

In analogy with the power spectrum in the Fourier series (see Sections 2.4 and 4.3), we have the quantity

$$| G(f) |^2 = G(f)\overline{G}(f)$$

as the power spectrum in the theory of Fourier integrals. Often the word "power" is dropped, and the quantity is merely referred to as the *spectrum*.

To obtain a feeling for what the Fourier integral is, we make the loose analogy of $g(t)$ as a light beam. The transform (8.2-1), like a glass prism, breaks up the function into its component frequencies f, each of intensity $G(f)$. In optics, the various frequencies are called colors: we get via the Fourier integral the color spectrum of the incoming signal. If the input is a single frequency, then we get a single *spectral line*. In practice, of course, an absolutely pure spectral line does not occur, but some, such as laser beams, are so close to a single frequency that the difference is often unimportant.

Unfortunately, if the function $g(t)$ does not approach zero as $|t|$ becomes infinite, then the transform $G(f)$ does not exist in a mathematical sense since the integrand (8.2-2) does not approach zero as t approaches infinity. This situation is a nuisance in the common case of

$$g(t) = \cos \omega_0 t$$

especially for $\omega_0 = 0_x$ for which $g(t) = 1$.

The difficulty is, however, a clear warning that the model being used is unrealistic in the following sense. We can truncate $g(t)$ by using

$$g_1(t) = \begin{cases} \cos \omega_0 t, & |t| < T \\ 0, & |t| > T \end{cases}$$

for some suitably large T and then carry out the desired work (involving the extra details due to the ends). If the discontinuities at the ends are annoying, we can put $g_1(t)$ through the rectangular convolving window and obtain a new $g_2(t)$ that has continuity and hence a more rapidly converging Fourier series. Indeed, any degree of smoothness at the ends can be used at the price of more details in processing the function. We then carry out the process for an arbitrary T. Finally let $T \to \infty$. If we obtain a peculiar result, then consider the following observation. These finite functions have finite energy. The fact that their energy becomes infinite as T becomes infinite shows the unrealistic nature of the model that assumes that they describe reality. A pure sinusoid lasting infinitely long is unrealistic, and some suitably truncated function should be practical in almost all situations.

The rest of the chapter simply describes some of the details of the preceding outline, and, as noted, we do not intend to be too rigorous. We will suppose that the functions used to represent the subject being studied are sufficiently "well behaved" to meet any special restraints on the class of functions for which the results apply and will exclude the "pathological cases." We are now definitely committed to the complex notation. For those whose mathematical experience with complex numbers is limited, it is suggested that the subject be reviewed before continuing; otherwise, the discussion will be obscured by the lack of elementary manipulative skills.

8.4 THE FOURIER INTEGRAL

We now derive the Fourier integral from the Fourier series. Given the Fourier series of a function that is periodic in the interval $-N < t \leq N$ (Section 4.9), we eliminate the coefficient c_k in the complex form and obtain

$$g(t) = \sum_{k=-\infty}^{\infty} \left[\int_{-N}^{N} g(t')e^{-(\pi i/N)\,kt'}\,dt' \right] e^{(\pi i/N)\,kt}\,\frac{1}{2N} \qquad (8.4\text{-}1)$$

where the dummy variable of integration is t' to avoid confusion. To approach the approximation of nonperiodic functions, we let the interval get longer and longer; that is, we let $N \to \infty$. In the limit the function $g(t)$ is no longer periodic, since the interval of periodicity is the whole axis.

To study in detail what happens as $N \to \infty$ we set

$$\frac{1}{2N} = \Delta f \qquad (8.4\text{-}2)$$

and hence define f_k, the kth point from the origin in either direction, by

$$\frac{k}{2N} = k\,\Delta f = f_k \qquad (8.4\text{-}3)$$

Our equation then becomes ($k = 2Nf_k$)

$$g(t) = \sum_{k=-\infty}^{\infty} \left[\int_{-N}^{N} g(t')e^{-2\pi i f_k t'}\, dt' \right] e^{2\pi i f_k t}\, \Delta f$$

It is easy to see from (8.4-2) and (8.4-3) that as N grows larger and larger, the successive f_k in the summation get closer and closer to each other; the exponentials are being pushed closer together, and the sum is approaching an integral. It is reasonable to suppose that in the limit the summation becomes an integral, *provided* the function $g(t)$ is reasonably well behaved. In the limit the equation becomes

$$g(t) = \int_{-\infty}^{\infty} \left[\int_{-\infty}^{\infty} g(t')e^{-2\pi i f t'}\, dt' \right] e^{2\pi i f t}\, df$$

To get the Fourier integral in the usual form, we simply set

$$G(f) = \int_{-\infty}^{\infty} g(t')e^{-2\pi i f t'}\, dt'$$

and we have (8.4-2) and (8.4-1)

$$g(t) = \int_{-\infty}^{\infty} G(f)e^{2\pi i f t}\, df$$

The use of the rotational frequency f rather than the angular frequency ω produces a nice symmetric representation without any extra numerical coefficients. The function $G(f)$ is said to be *the Fourier transform* of the function $g(t)$. The two functions have almost exactly the same relationship to each other; the exception is that in the exponent one integral has i and the other $-i$. Both functions contain the same information in the sense that each can be found from the other; but *they exhibit the information in significantly different forms*. It is the power of these alternate forms that makes the Fourier transform so useful in understanding what is occurring in many situations.

8.5 SOME TRANSFORM PAIRS

The importance of the relationship between the two functions $g(t)$ and $G(f)$ suggests that a table of transforms would be useful. Such tables are widely available. For our immediate use, we will develop only a few such relationships that we will need, and leave the more extensive development of the Fourier integral to books that specialize in the subject.

Our first example of a Fourier integral transform is the band-limited function with unit area [Figure 8.5-1(a)] [we are effectively repeating (8.3-6)]

$$G(f) = \begin{cases} \dfrac{1}{2f_c}, & \text{for } |f| < f_c \\[2mm] 0, & \text{for } |f| > f_c \end{cases} \tag{8.5-1}$$

We now set up the integral for $g(t)$ and, as in Section 8.3, obtain

$$\begin{aligned}
g(t) &= \int_{-\infty}^{\infty} G(f)e^{2\pi ift}\, df = \frac{1}{2f_c}\int_{-f_c}^{f_c} e^{2\pi ift}\, df \\[2mm]
&= \frac{1}{2f_c}\frac{e^{2\pi ift}}{2\pi it}\bigg|_{-f_c}^{f_c} = \frac{e^{2\pi if_ct} - e^{-2\pi if_ct}}{2i}\frac{1}{2f_c\pi t} \\[2mm]
&= \frac{\sin 2\pi f_c t}{2\pi f_c t}
\end{aligned} \tag{8.5-2}$$

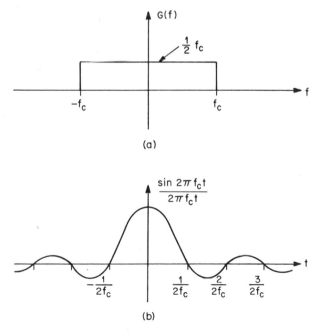

(a)

(b)

FIGURE 8.5-1

This function is very common in the theory, and an idea of its behavior around the origin is given by its power series expansion

$$\frac{\sin 2\pi f_c t}{2\pi f_c t} = 1 - \frac{(2\pi f_c t)^2}{3!} + \frac{(2\pi f_c t)^4}{5!} - \cdots$$

Figure 8.5-1(b) gives the graph of the function. The main lobe extends from $-1/(2f_c)$ to $1/(2f_c)$; beyond that the function has equally spaced zeros, while the amplitude gradually dies out like $1/t$. The larger the f_c, the narrower the central peak and the spacing of the zeros.

Corresponding to the rectangular window for the Fourier series we have been using one for the Fourier integral. The use of the unit impulse function $\delta(x)$ leads to the peculiar mathematical expressions

$$\int_{-\infty}^{\infty} \delta(x) \, dx = 1 \quad \text{and} \quad \int_{-\infty}^{\infty} f(x)\delta(x - k) \, dx = f(k)$$

Thus we have unit step and ramp functions for continuous variables to work with the Fourier integral. Notice that we have not, and in general will not, use them because if you look first at either the summation or the integral then the effect is usually merely to limit the range (or to select a value). The special functions of the z-transform are needed mainly when you look first at the quantity being summed, or else the integrand, and then impose the limits before the summation or the integration. This is not to say that they are not at times useful, but that in our simple problems the special notations only serve to hide the essentials of what is going on.

Our second example of a Fourier transform is actually a relationship between Fourier transforms. Suppose that we already know a pair of Fourier transforms $g_1(t)$ and $G_1(f)$. Thus we assume that we know

$$g_1(t) = \int_{-\infty}^{\infty} G_1(f)e^{2\pi i f t} \, df$$

We now ask: What function $g_2(t)$ corresponds to $G_1(f)e^{2\pi i f y}$? From the definition, we have

$$g_2(t) = \int_{-\infty}^{\infty} [G_1(f)e^{2\pi i f y}]e^{2\pi i f t} \, df$$

$$= \int_{-\infty}^{\infty} G_1(f)e^{2\pi i f(y + t)} \, df$$

Clearly, $y + t$ plays the role of t in the original function, so that

$$g_z(t) = g_1(y + t) \tag{8.5-3}$$

The effect of the exponential multiplier $\exp(2\pi i f y)$ is to shift the argument of the transformed function. This result is often called the *shifting theorem*.

Exercises

8.5-1   Find the Fourier transform of $g(t) = \begin{cases} |t|, & |t| < t_0 \\ 0, & \text{otherwise} \end{cases}$

8.5-2   Find the Fourier transform of $g(t) = \begin{cases} 1 - |t|, & |t| < 1 \\ 0, & |t| > 1 \end{cases}$

8.5-3   If $g(t) = (1/\sqrt{2\pi})e^{-at^2}$, find its Fourier transform.

8.5-4   Find the Fourier transform of $e^{-a|t|}$.

8.6   BAND-LIMITED FUNCTIONS AND THE SAMPLING THEOREM

Because of the importance of the sampling theorem in computing, we present a second, slightly more rigorous proof in this section. We are still avoiding pathological functions and excessive rigor.

The central idea of the sampling theorem is that a band-limited function $g(t)$ extending from $t = -\infty$ to $t = \infty$ is sampled at equally spaced points with a spacing such that at least two samples per cycle occur in the highest frequency present. As noted, the restriction to band-limited functions corresponds to a natural physical limitation in many situations. In any case, it is forced on us if we sample a function, since the very act of sampling produces the aliasing of the higher frequencies into the frequencies within the Nyquist band. However, a word of caution: the mathematical model being used asserts that, if a signal is band limited, then it cannot be time limited; and, conversely, if it is time limited, then it cannot be band limited. In practice, no signal lasts forever, and so in the mathematical model it cannot be band limited. Also there are upper limits to possible frequencies. Evidently, the mathematical model should not be taken too literally when applying results to the real world; it is a useful, not a sacrosanct model.

To derive the sampling theorem, suppose that we are given the band-limited function $G(f)$, which is zero for $|f| > f_c$. The first step in the derivation is to replace this function with a periodic function (in order to use our theory of Fourier series) that agrees with $G(f)$ in the band; thus we define the function $G_1(f)$, which coincides with $G(f)$ inside the interval $|f| < f_c$ and is periodic outside, as shown in Figure 8.6-1.

Using the Fourier integral, we have

$$G(f) = \int_{-\infty}^{\infty} g(t)e^{-2\pi i f t}\, dt \qquad (8.6\text{-}1)$$

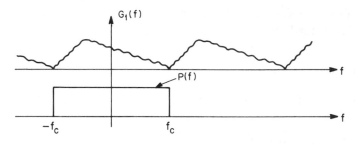

FIGURE 8.6-1

Since $G(f)$ is band limited, and in the band we have $G(f) \equiv G_1(f)$,

$$g(t) = \int_{-\infty}^{\infty} G(f)e^{2\pi ift} \, df = \int_{-f_c}^{f_c} G(f)e^{2\pi ift} \, df = \int_{-f_c}^{f_c} G_1(f)e^{2\pi ift} \, df$$

$$(8.6\text{-}2)$$

For $G_1(f)$ we have, since we made it periodic, the Fourier series expansion

$$G_1(f) = \sum_{k=-\infty}^{\infty} c_k e^{(\pi i/f_c)kf} \qquad (8.6\text{-}3)$$

where from (4.9-8) modified slightly

$$c_k = \frac{1}{2f_c} \int_{-f_c}^{f_c} G_1(f)e^{-(\pi i/f_c)kf} \, df \qquad (8.6\text{-}4)$$

But from the equation (8.6-4) for $g(t)$ this integral is by comparing exponents the same as

$$c_k = \frac{1}{2f_c} g\left(\frac{-k}{2f_c}\right)$$

To get the original function $G(f)$ from the new periodic function $G_1(f)$, we multiply $G_1(f)$ by the rectangular function $2f_c P(f)$, (we "cut out" the function) where (see Figure 8.5-1 again)

$$P(f) = \begin{cases} \dfrac{1}{2f_c}, & \text{for } |f| < f_c \\[2ex] 0, & \text{for } |f| > f_c \end{cases}$$

We have already found that the transform of $P(f)$ (8.5-2) is the band-limited function

$$p(t) = \frac{\sin 2\pi f_c t}{2\pi f_c t}$$

Thus to get the factor of 1 we must multiply by $2f_c$

$$G(f) = G_1(f)P(f)2f_c \qquad (8.6\text{-}5)$$

For $G_1(f)$ and the c_k, we substitute the results just found (8.6-3) and (8.6-4), and we have from (8.5-5)

$$G(f) = \sum_{k=-\infty}^{\infty} g\left(\frac{-k}{2f_c}\right) P(f) e^{(\pi i / f_c) k f} \qquad (8.6\text{-}6)$$

The next step is to transform the results back to the time domain via the Fourier transform and apply the shifting theorem (8.5-3) to (8.6-6)

$$g(t) = \int_{-\infty}^{\infty} G(f) e^{2\pi i f t} \, df$$

$$= \sum_{k=-\infty}^{\infty} g\left(\frac{-k}{2f_c}\right) \int_{-\infty}^{\infty} P(f) e^{\pi i k f / f_c} e^{2\pi i f t} \, df$$

$$= \sum_{k=-\infty}^{\infty} g\left(\frac{-k}{2f_c}\right) \frac{\sin 2\pi f_c(t + k/2f_c)}{2\pi f_c(t + k/2f_c)}$$

Now replacing k by $-k$ as the dummy summation index, we have the sampling theorem

$$g(t) = \sum_{k=-\infty}^{\infty} g\left(\frac{k}{2f_c}\right) \frac{\sin \pi(2f_c t - k)}{\pi(2f_c t - k)} \qquad (8.6\text{-}7)$$

At the folding frequency itself we cannot reconstruct the function from the samples because, for unit spacing, the function $g(t) = \sin \pi t$ is identically zero at all the sample points, and the sampling theorem would therefore give the identically zero function. Thus, in the "splitting hairs" mathematical approach, we must have *more* than two samples per cycle for the highest frequency present.

8.7   THE CONVOLUTION THEOREM

For our purposes, perhaps the most important theorem in the theory of Fourier integrals (after the inversion formula) is the *convolution theorem*, which corresponds to the earlier convolution theorems for Fourier series (Section 5.2). Suppose that we have two functions, $g_1(t)$ and $g_2(t)$. The convolution $h(t)$ of $g_1(t)$ with $g_2(t)$ is defined as

$$h(t) = \int_{-\infty}^{\infty} g_1(s)g_2(t - s) \, ds \tag{8.7-1}$$

(Notice how it differs from the earlier definition in Section 5.2, where we assumed periodicity of the functions.) The dummy variable of integration is s; and regarding t as fixed with respect to the integration, we make the change of variable of integration

$$t - s = s'$$

The convolution becomes

$$h(t) = \int_{-\infty}^{\infty} g_1(t - s')g_2(s') \, ds'$$

So (as before, Section 5.2) the convolution of $g_1(t)$ with $g_2(t)$ is the same as the convolution of $g_2(t)$ with $g_1(t)$.

It is natural to ask: What is the Fourier transform of the convolution? By definition, the Fourier transform of $h(t)$ is

$$H(f) = \int_{-\infty}^{\infty} h(t)e^{-2\pi ift} \, dt$$

Substituting for $h(t)$, we get [recognizing the square bracket as $G_2(f)$]

$$H(f) = \int_{-\infty}^{\infty} g_1(s)e^{-2\pi ifs} \left[\int_{-\infty}^{\infty} g_2(t - s)e^{-2\pi i(t-s)f} \, dt \right] ds$$

$$= \int_{-\infty}^{\infty} g_1(s)e^{-2\pi ifs} G_2(f) \, ds$$

$$= G_1(f)G_2(f) \tag{8.7-2}$$

Thus we have the important result that *the Fourier transform of a convolution of two functions is the product of their Fourier transforms.*

The preceding result has been shown in the time domain, but, by symmetry of the transforms, it applies equally in the frequency domain.

Occasionally, the convolution of a function with respect to itself is of some interest. We have

$$h(t) = \int_{-\infty}^{\infty} g(s)g(t - s) \, ds$$

Applying the Fourier transform twice gives

$$h(t) = \int_{-\infty}^{\infty} G^2(f)e^{-2\pi ift} \, df$$

and for $t = 0$ we have

$$h(0) = \int_{-\infty}^{\infty} g(s)g(-s) \, ds = \int_{-\infty}^{\infty} G^2(f) \, df$$

Written in the form

$$\int_{-\infty}^{\infty} |g(s)|^2 \, ds = \int_{-\infty}^{\infty} |G(f)|^2 \, df \tag{8.7-3}$$

it is known as *Parseval's theorem* and corresponds to (4.9-5) as $N \to \infty$.

Exercises

8.7-1 Fill in the details for the second form of the convolution theorem.

8.7-2 Fill in the details for Parseval's theorem.

8.8 THE EFFECT OF A FINITE SAMPLE SIZE

Very often the time function $g(t)$ can be regarded as an infinitely long signal (in time). Of necessity, we must take a finite-length sample. Thus in astronomy we can observe a pulsar or a Cepheid variable for a finite length of time only. What does this limitation do to the original signal? We can regard the limitation as being equivalent to multiplication in time by the rectangular pulse of unit height $2Tp(t)$ (which is a multiplying window), where

$$p(t) = \begin{cases} \dfrac{1}{2T}, & |t| < T \\ 0, & |t| > T \end{cases}$$

This is also known as the "box car function," "gate function," and the "cookie cutter function." Thus what we see is not $g(t)$ but rather $g_1(t)$, defined by (see Section 8.5)

$$g_1(t) = 2Tg(t)p(t)$$

The Fourier transform is, by the convolution theorem,

$$G_1(f) = 2T \int_{-\infty}^{\infty} G(f_1) \frac{\sin 2\pi T(f_1 - f)}{2\pi T(f_1 - f)} df_1$$

where f_1 is the dummy variable of integration. So the transform of what we see is the convolution of the true transform $G(f_1)$ with the convolving window (f_1 is the variable)

$$2T \frac{\sin 2\pi T(f_1 - f)}{2\pi T(f_1 - f)}$$

In the situation in which the original signal is a single frequency (a spike or line in the spectrum), this line is convolved with the above sinusoidal function, and therefore we will see this same function, the Gibbs phenomenon, in a slightly new form. It is the usual $(\sin f)/f$ function with the amplitude of the main lobe growing in height like $2T$ and the half-width to the nearest zero contracting like $1/2T$.

If we have many lines in the spectrum, instead of a single line, each will be smeared out by the same $(\sin f)/f$-type function, and in the limit of a continuous spectrum we will get the foregoing convolution integral. In an optical analogy used earlier, this smearing out corresponds to the idea of resolving power: the longer the time signal is observed, the better we can resolve (separate) adjacent spectral lines. Although intuitively evident that something similar should occur, we have exhibited the actual dependence on T, where $2T$ is the length of the observation. Thus we can now understand what happens to a function when we chop out a piece of it to observe closely. The length of the sample we take limits what we can hope to discover from the sample.

If there are two lines, or other features, in the spectrum that are close to each other, then unless we have a long run of observations, the convolving window will not let us distinguish them from a single line. The closer the two lines are to each other, the longer the length of the observation must be if we are to resolve them. In a sense, the situation is similar to two competing teams. The closer the teams are in ability, the longer we must observe them in order to decide which team is the better.

This result, that "chopping out" a segment from a continuous function amounts to convolving the transform of the function with

$$2T\,\frac{\sin 2\pi T(f_1 - f)}{2\pi T(f_1 - f)}$$

should be compared with the earllier "chopping out" of $2N + 1$ terms from a Fourier series in Section 5.5, where we obtained the corresponding convolving function

$$h(t) = \frac{\sin(N + 1/2)t}{\sin(t/2)}$$

and the modified window

$$h(t) = \frac{\sin Nt \cos t/2}{\sin t/2}$$

The comparison (allowing for the trivial fact that one is in the frequency variable and one is in the time variable) shows the relationship of the continuous model of the original signal and the discrete sampled model used in the actual data processing. In comparing the discrete and continuous rectangular windows we see denominators of $\sin t/2$ and $t/2$.

8.9 WINDOWS

Now we have the formal apparatus of the Fourier integral, the subject of windows is a little clearer. In the domain of continuous functions (and not just transfer functions), the convolution theorem shows what happens to the continuous functions that we tend to hold in our minds; it is the discrete-continuous convolution theorems that tell us what occurs in the actual case being dealt with, that is, the discrete samples of the function. For both cases, windowing in one domain amounts to multiplication by the transform of the window, discrete or continuous, in the other domain. And it is inevitable; we can do nothing about it. The chopping out of a sample from an infinitely long function is windowing. Window theory tells us what to expect, as well as what to do rather than use a rectangular window if we want to reduce the effects of the discontinuities.

The use of a shaped window like the raised cosine of Section 5.7 has a similar result in the theory of Fourier integrals. The Fourier transform of

$$(2T)h(t) = \begin{cases} \dfrac{1 + \cos \pi t/T}{2}, & |t| < T \\ 0, & |t| > T \end{cases}$$

is

$$
\begin{aligned}
H(f) &= \frac{1}{2T} \int_{-T}^{T} \frac{1 + \cos \pi t/T}{2} e^{-2\pi ift} \, dt \\
&= \frac{1}{4T} \int_{-T}^{T} [e^{\pi it/T} + 2 + e^{-\pi it/T}] e^{-2\pi ift} \, dt \\
&= \frac{1}{4} \frac{\sin 2\pi T(f - 1/2T)}{2\pi T(f - 1/2T)} + \frac{1}{2} \frac{\sin 2\pi fT}{2\pi fT} + \frac{1}{4} \frac{\sin 2\pi T(f + 1/2T)}{2\pi T(f + 1/2T)} \\
&= \frac{1}{4} \frac{\sin 2\pi(fT - 1/2)}{2\pi(fT - 1/2)} + \frac{1}{2} \frac{\sin 2\pi fT}{2\pi fT} + \frac{1}{4} \frac{\sin 2\pi(fT + 1/2)}{2\pi(fT + 1/2)}
\end{aligned}
$$

which can be rearranged by expanding the sines of the sums in both end numerators ($\sin \pi = 0$, $\cos \pi = -1$):

$$H(f) = \frac{1}{4} \frac{\sin 2\pi fT}{\pi} \left[-\frac{1}{2fT - 1} + \frac{2}{2fT} - \frac{1}{2fT + 1} \right]$$

Additional algebra results in the form analogous to the one in Section 5.7:

$$H(f) = \frac{1}{2} \frac{\sin 2\pi fT}{2\pi fT} \left[\frac{1}{1 - (2fT)^2} \right]$$

The Hamming window in the continuous case may similarly be derived by following the method of Section 5.8. Alternately, we may note that the operations of smoothing and sampling may be interchanged (except for end effects at most) and deduce that the corresponding continuous case will have the appropriate form. We can see by Figure 5.8-1 that the values of the weights in the limiting case approach the continuous case values, 0.23, 0.54, 0.23, as they should.

A comparison of the discrete Fourier series and the continuous Fourier integral results shows that many times where the Fourier integral has a denominator that is linear in the variable (for example, immediately above $2\pi fT$) the corresponding formula for the series will have a sine term. This arises from the aliasing that would occur owing to the sampling to get from the continuous to the discrete (periodic) representation.

Exercise

8.9-1   Make a table of correspondences between the Fourier series formulas and those of the Fourier integral.

8.10   THE UNCERTAINTY PRINCIPLE

There is an interesting relationship between a function and its Fourier transform, which we now derive. It is not needed directly in this text, but it gives some insight into the question of how "spread out" a function and its transform must be.

We have seen in one case (Section 8.5) that as one function gets narrower and narrower the transform gets wider and wider. In Section 1.5 we introduced the concept of the *variance* as a measure of how "spread out" a distribution is. It is no loss of generality to assume that the mean of the distribution is at the origin; therefore, we measure the variance of a function (distribution) $h(t)$ by

$$\text{Var }\{h(t)\} = \int_{-\infty}^{\infty} t^2 \mid h(t) \mid^2 dt \qquad (8.10\text{-}1)$$

We will scale any function we use to have the normalization

$$\int_{-\infty}^{\infty} \mid h(t) \mid^2 dt = 1 \qquad (8.10\text{-}2)$$

and by (8.7-3) we see that we also have

$$\int_{-\infty}^{\infty} \mid H(f) \mid^2 df = 1$$

We also impose the condition that as $\mid t \mid \to \infty$

$$\sqrt{t}\, h(t) \to 0 \qquad (8.10\text{-}3)$$

Corresponding to the spread of $h(t)$, we have the spread of its transform,

$$\text{Var }\{H(f)\} = \int_{-\infty}^{\infty} f^2 \mid H(f) \mid^2 df$$

Again we are assuming that the mean is at zero (since a shift amounts to a pure imaginary exponent and this cancels out when we use the absolute value).

The uncertainty principle we will prove states that

$$\text{Var} \{h(t)\} \, \text{Var} \{H(f)\} \geq \frac{1}{16\pi^2} \tag{8.10-4}$$

To prove this we need the famous Schwarz inequality, which is easy to derive. Let $g_1(t)$ and $g_2(t)$ be any two functions satisfying (8.10-3), and consider as a function of x

$$\int_{-\infty}^{\infty} |\, xg_1(t) + g_2(t) \,|^2 \, dt \geq 0$$

$$x^2 \int_{-\infty}^{\infty} |\, g_1(t) \,|^2 \, dt + 2x \int_{-\infty}^{\infty} |\, g_1(t)g_2(t) \,| \, dt + \int_{-\infty}^{\infty} |\, g_2(t) \,|^2 \, dt \geq 0$$

This is a quadratic equation of the form

$$Ax^2 + 2Bx + C \geq 0$$

This quadratic in x cannot have two real distinct zeros since, if there were, the parabola would be negative somewhere (it could have, of course, a double zero). The condition for this is

$$B^2 - AC \leq 0$$

Substitute for A, B, and C to get (use triangle inequality on left)

$$\left| \int_{-\infty}^{\infty} g_1(t)g_2(t) \, dt \right| \leq \left(\int_{-\infty}^{\infty} |\, g_1(t)g_2(t) \,| \, dt \right)^2$$

$$\leq \int_{-\infty}^{\infty} |\, g_1(t) \,|^2 \, dt \int_{-\infty}^{\infty} |\, g_2(t) \,|^2 \, dt \tag{8.10-5}$$

This is the Schwarz inequality in the form we need it.

We now define the functions $g_1(t)$ and $g_2(t)$ as ($h(t)$ real)

$$g_1(t) = th(t)$$

$$g_2(t) = \frac{dh(t)}{dt}$$

where $h(t)$ obeys equations (8.10-2) and (8.10-3). On the left we have

$$\int_{-\infty}^{\infty} th(t) \frac{dh(t)}{dt} \, dt = t \frac{h^2(t)}{2} \Big|_{-\infty}^{\infty} - \frac{1}{2} \int_{-\infty}^{\infty} h^2(t) \, dt = -\frac{1}{2}$$

Next,

$$h(t) \Leftrightarrow H(f)$$

$$\frac{dh(t)}{dt} \Leftrightarrow 2\pi i f H(f)$$

Hence (8.10-5) becomes

$$\frac{1}{4} \leq \int_{-\infty}^{\infty} t^2 \mid h(t) \mid^2 dt \int_{-\infty}^{\infty} 4\pi^2 f^2 \mid H(f) \mid^2 df$$

from which (8.10-4) follows immediately.

This is the famous uncertainty principle of quantum mechanics in the units we are using. In physics there is the identification of "conjugate variables" with the corresponding Fourier transforms, but due to scaling the constant in the inequality has a different value (involving Planck's constant).

It is worth investigating when the equality can hold. This requires that

$$xg_1(t) + g_2(t) = 0 \qquad \text{for all } t$$

Thus $g_1(t)$ is proportional to $g_2(t)$, and changing notation to a more conventional form we have

$$\lambda g_1(t) + g_2(t) = 0$$

$$\lambda th(t) + h'(t) = 0$$

$$\frac{dh}{h} = -\lambda t \, dt$$

$$\ln h = \frac{-\lambda t^2}{2} + C_1$$

$$h = C_2 e^{-\lambda t^2/2}$$

This is a gaussian distribution (1.5-2), and the constant C_2 is determined by the condition (8.10-2). In other words this $g(t)$ is the narrowest one you can get in quantum mechanics in the sense of the smallest product of variances.

This is worth a little reflection, especially since the uncertainty principle is so highly touted in quantum mechanics. We saw in Chapter 2 that, if we have a linear, time-invariant system, then the eigenfunctions are

$$e^{i\omega t} \equiv e^{2\pi i f t}$$

and these are the "natural functions" to use. When, therefore, we want to represent a general function we are forced to the Fourier transform representation of the function

$$h(t) = \int_{-\infty}^{\infty} H(f)e^{2\pi i f t}\, df$$

Thus we conclude that *any linear, time-invariant system must have an uncertainty principle* involving Fourier transforms. "A function and its transform cannot both be narrow" is what this section is showing, and this agrees with section 8.5.

9

Kaiser Windows
and Optimization

9.1 WINDOWS

The purpose of this chapter is to improve on the earlier nonrecursive filter
design methods of Chapters 3, 6, and 7. Before doing so, however, let us
review where we are and where we have been in order to clarify what we
are doing. We began with digital filters, which are linear combinations of
the data. Behind the design of filters is the data that is to be processed by
them. We are still looking at nonrecursive filters, which use only the data
and not the values that have been computed from the data. Furthermore,
we are looking at the two special cases of symmetric (even) and skew-sym-
metric (odd) filters and a linear combination of them can represent the general case of arbitrary coefficients. If the original signal is

$$u(t)$$

we sample and quantize this function at a uniform spacing of unit time. Thus
we actually observe the equally spaced data

$$u_n \qquad (9.1\text{-}1)$$

and may have caused aliasing by the act of sampling.

The u_n are then processed by a nonrecursive time-invariant filter of the
form (the coefficients c_k do not depend on n)

$$y_n = \sum_{k=-\infty}^{k=\infty} c_k u_{n-k}$$

185

In principle (Figure 6.1-4), there can be an infinite number of terms in the Fourier expansion, but, in practice, the sum runs between finite limits

$$y_n = \sum_{k=-N}^{k=N} c_k u_{n-k} \tag{9.1-2}$$

Since the system is linear, it follows that an eigenfunction

$$u(t) = Ae^{i\omega t} = Ae^{2\pi i f t}$$

will produce the output from the filter

$$y_n = \tilde{H}(f)Ae^{2\pi i f n}$$

where $\tilde{H}(f)$ depends on the c_k (and, of course, on f but not on n). Thus, as noted before, the transfer function $\tilde{H}(f)$ is the corresponding eigenvalue. And a sum of such eigenfunctions

$$u_n = \sum_{k=1}^{N} A_k e^{2\pi i f_k n} \tag{9.1-3}$$

will give the corresponding sum at the output

$$y_n = \sum_{k=1}^{N} A_k \tilde{H}(f_k)e^{2\pi i f_k n}$$

Each eigenfunction "minds its own business."

Now that we have the theory of the Fourier integral, we know that any reasonable function $g(t)$ has a representation [a decomposition into frequencies of amounts $G(f)$] of the form

$$g(t) = \int_{-\infty}^{\infty} G(f)e^{2\pi i f t} \, df$$

The $G(f)$ corresponds to the A in the foregoing (9.1-3) eigenfunction discussion, so the output from the filter is

$$g_1(t) = \int_{-\infty}^{\infty} \tilde{H}(f)G(f)e^{2\pi i f t} \, df$$

For the class of nonrecursive filters of the smoothing, stopband, passband, and interpolation forms, we have

$$c_k = c_{-k}$$

And for them the transfer function

$$\tilde{H}(f) - \sum_{k=-N}^{N} c_k e^{2\pi i f k} \tag{9.1-4}$$

can also be written

$$\tilde{H}(f) = c_0 + 2 \sum_{k=1}^{N} c_k \cos 2\pi f k \tag{9.1-5}$$

The period of the transfer function is that of the Nyquist interval, and it arises from the sampling process with the corresponding aliasing. Thus, for practical purposes, the transfer function has meaning only in the interval (remember we are using a *unit* sampling rate)

$$-\tfrac{1}{2} \le f \le \tfrac{1}{2}$$

9.2 REVIEW OF GIBBS PHENOMENON AND THE RECTANGULAR WINDOW

Consider once more the low-pass filter and its transfer function. In Section 6.3 we found its Fourier series,

$$\tilde{H}(f) = 2f_s + 2 \sum_{k=1}^{k=\infty} \left[\frac{1}{\pi k} \sin 2\pi k f_s \right] \cos 2\pi k f$$

In practice, this series must be truncated to a finite number of terms. The process is equivalent to multiplying the terms of the Fourier series in its complex form by the corresponding terms of the series ($2N + 1$ nonzero terms)

$$0, 0, 1, 1, \ldots, 1, 0, 0$$

In Section 5.5 we showed that this procedure is equivalent to convolving the transfer function $\tilde{H}(f)$ with (notice that we change from t to $2\pi f_1$ as the independent variable)

$$W(f_1) = \frac{\sin \pi(2N + 1)f_1}{\sin \pi f_1}$$

Thus the transfer function represented by the truncated Fourier series is the *convolution* of the ideal transfer function with this ratio of two sines,

$$\tilde{H}_1(f) = \int_{-\infty}^{\infty} \tilde{H}(f_1)W(f - f_1)\,df_1$$

We picture this convolution of the transfer function $\tilde{H}(f_1)$ with the ratio of sinusoids $W(f_1)$ (which appears as a gradually dying out sinusoid starting at a middle arch that is twice as wide as the other intervals between zeros), and we watch as the tail approaches the edge at the frequency, $-f_s$ (Figure 9.2-1), of the low-pass filter being designed. We see that it is the ratio of the two sinusoids that is integrated by the rectangular shape. Thus we must think about the integral of the window times the sinusoid ratio. As the convolution moves across the interval, more and more of the sinusoid ratio comes into the rectangular pulse; and when the main arch enters, a great rise in the value of the integral occurs. Finally, as the convolution continues, we see the dying out of the ripples in the integral due to the first edge. There is, of course, also the second edge at f_s of the filter to be considered. This figure displays the Gibbs phenomenon exactly; we see it in a new light. When we remember that both the ratio of the sinusoids and the original filter shape are periodic, we have a complete and accurate understanding of how the Gibbs phenomenon occurs as a result of truncation of the original series.

Recall that in Section 5.4 we looked at Lanczos' suggestion of how to reduce the Gibbs effect due to truncation. He suggested convolving the orig-

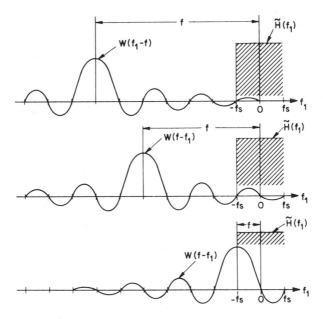

FIGURE 9.2-1 GIBBS PHENOMENON

inal rectangular function with a square window whose width is adjusted to the length of the ripple in the Gibbs wiggles. This process, we found, was equivalent to multiplying the coefficients of the Fourier series by the sigma factors, and it gave a reduction of almost a factor of 10 in the ripple size.

A repetition of the Lanczos smoothing was seen to be equivalent to a triangular window of twice the width, and it produced the square of the sigma factor, thereby making the effect of the higher frequencies even less, but also making the transition band twice as wide. We also looked at von Hann and Hamming windows.

This discussion suggests that we look further at what shape of window to use when truncating the Fourier series.

9.3 THE KAISER WINDOW: I_0-sinh WINDOW

What do we want for a window? We would like both the window and its transform to be narrow. But this situation is impossible in reality (see Section 8.10). How shall we compromise? If we are in the domain of continuous variables, then the *prolate spheroidal functions* are known to have the property that in a certain sense both are limited as much as possible. However, we are in the discrete domain of Fourier coefficients and have seen that there is a difference in the two cases. Reasoning that the difference between the two is not great in practice and that only a good approximation is needed to do the job, J. F. Kaiser suggested that we use the weights on the Fourier coefficients

$$w(k) = \begin{cases} \dfrac{I_0[\alpha\sqrt{1 - (k/N)^2}]}{I_0(\alpha)}, & |k| \le N \\ 0, & |k| > N \end{cases} \qquad (9.3\text{-}1)$$

in place of the sigma factors of the rectangular window, where

$$I_0(x) = 1 + \sum_{n=1}^{\infty} \left[\frac{(x/2)^n}{n!} \right]^2 \qquad (9.3\text{-}2)$$

For the knowledgeable the $I_0(x)$ is the Bessel function of order zero and pure imaginary argument. We need not be intimidated by these words, it will simply be a subroutine which we will carefully describe. See Figure 9.3-1 for a sketch of $I_0(x)$.

Notice how Kaiser's weights resemble the Hamming raised cosine on

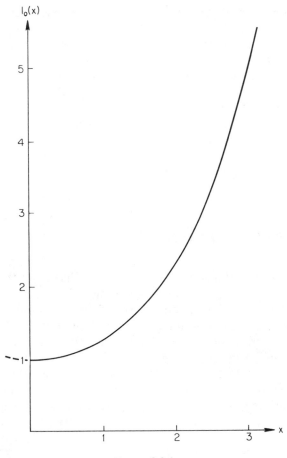

FIGURE 9.3-1

a platform, since at the ends of the nonzero terms we have the value $w(N)$ $= 1/I_0(\alpha)$, whereas at the middle we have the value $w(0) = 1$. The parameter α thus determines the height of the platform.

Kaiser's weights contain two parameters: N, which is the half-width of the window (where we keep $2N + 1$ complex Fourier coefficients), and α, which controls the "shape" of the window, in particular, how large the ripples will be.

It can be shown that the transform of the $w(k)$, now regarded as a continuous function, is the function

$$W(f) = \frac{2N \sinh[\alpha\sqrt{1 - (f/f_a)^2}]}{\alpha I_0(\alpha)\sqrt{1 - (f/f_a)^2}} \qquad (9.3\text{-}3)$$

where

$$f_a = \frac{\alpha}{N}$$

when $f > f_a$ then we have

$$\sqrt{1 - (f/f_a)^2} = i\sqrt{(f/f_a)^2 - 1}$$

But

$$\sinh x = \frac{e^x - e^{-x}}{2}$$

hence

$$\sinh ix = \frac{e^{ix} - e^{-ix}}{2} = i \sin x$$

The radical in the denominator of (9.3-3) also produces a factor i and hence

$$W(f) = \frac{2N \sin[\alpha\sqrt{(f/f_a)^2 - 1}]}{\alpha I_0(\alpha)\sqrt{(f/f_a)^2 - 1}} \qquad |f| > f_a \qquad (9.3\text{-}4)$$

At $f = f_a$ we use L'Hôpital's rule and have

$$W(f_a) = \frac{2N}{I_0(\alpha)}$$

and is sinusoidal with decaying ripples (due to the increasing size of the denominator) of amplitude approximately $1/f$; see Figure 9.3-2. The zeros approach equal spacing as $f \to \infty$.

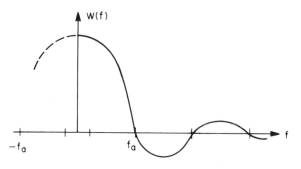

FIGURE 9.3-2 KAISER WINDOW

TABLE 9.3-1 Performance of I_0-sinh
window for equally spaced values of the
attenuation

A(dB)	α	D
25	1.333	1.187
30	2.117	1.536
35	2.783	1.884
40	3.395	2.232
45	3.975	2.580
50	4.551	2.928
55	5.102	3.276
60	5.653	3.625
65	6.204	3.973
70	6.755	4.321
75	7.306	4.669
80	7.857	5.017
85	8.408	5.366
90	8.959	5.714
95	9.501	6.062
100	10.061	6.410

In the transform domain, $W(f)$ of Figure 9.3-2 is convolved with the rectangular shape of the ideal filter; compare with Figure 9.2-1. Thus it is the convolution of the transform $W(f)$ that makes the ripples in the frequency domain. The maximum allowed ripple will be called δ in both the stopband and passband.

This convolution was computed numerically by Kaiser. Table 9.3-1 gives a quantity α related to the maximum overshoot as a function of attenuation A (in decibels), where we will explain the symbols in detail in the following example. We note for future reference that for $\alpha = 0$, $W(f)$ can be written

$$W(f) = \frac{2N}{I_0(\alpha)} \left[\frac{\sinh[\alpha\sqrt{1 - (f/f_a)^2}]}{\alpha\sqrt{1 - (f/f_a)^2}} \right]$$

and since $I_0(\alpha) = 1$, we have at $\alpha = 0$

$$W(f) = 2N$$

Thus we have the pure case of Gibbs with no window shaping at all. The $\alpha = 5.4414$ corresponds (has the same mainlobe width) to the Hamming window, while $\alpha = 8.885$ similarly corresponds to the Blackman window [BT, p. 98]. The third column of the table gives the window width D.

Before becoming enmeshed in the details of how the formulas occur, we will first go through the steps of the design of a filter.

We begin by making a sketch of the ideal filter (Figure 9.3-3) that we

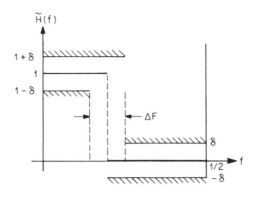

FIGURE 9.3-3 IDEAL FILTER

would like to have. This figure includes the acceptable width ΔF of the transition band (bands) and δ, the ripple half-amplitude that we can tolerate.

Second, we calculate the attenuation A in decibels:

$$A = -20 \log_{10} \delta \qquad (9.3\text{-}5)$$

From the attenuation A we find the α to shape the window [because the tails of $W(f)$ produce the ripples in the convolution]. To do so, we use the empirical formula that Kaiser derived by curve fitting Table 9.3-1.

$$\alpha = \begin{cases} 0.1102(A - 8.7), & 50 < A \\ 0.5842(A - 21)^{0.4} + 0.07886(A - 21), & 21 < A < 50 \\ 0, & A \le 21 \end{cases} \qquad (9.3\text{-}6)$$

A sketch of this function appears in Figure 9.3-4. In this figure we see that the Gibbs 8.9% overshoot (21-dB attenuation) corresponds to the value $A = 21$.

Finally, the number of terms to keep, $(2N + 1)$, is given by the formula for N,

$$N \ge \frac{A - 7.95}{28.72 \, \Delta F} \qquad (9.3\text{-}7)$$

This formula shows that N is inversely proportional to the width of the transition band ΔF. The attenuation A depends on the log of the ripple height δ, and thus it is much easier to reduce the ripple size than it is to narrow the transition band, as judged by the number of terms in the final filter. In this way, we find $w(k)$ and $W(f)$ for the windowing of the original Fourier series approximation.

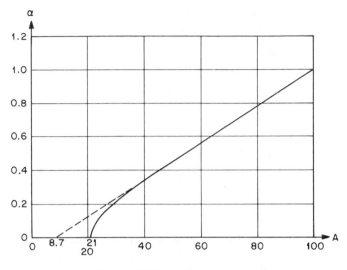

FIGURE 9.3-4 α versus Attenuation

9.4 DERIVATION OF THE KAISER FORMULAS

We begin the derivation of the formulas by recalling that the span of a nonrecursive filter of $2N + 1$ terms is $2N$ intervals and that we used the normalized frequency so that the Nyquist interval reaches from $-\frac{1}{2}$ to $\frac{1}{2}$. The normalized transition width is ΔF.

Kaiser noted that, for a fixed δ, the product of the two widths, the window span and the transition width, was a constant; it was defined as the D factor

$$(2N)(\Delta F) = D$$

It is now necessary to find a formula for N that gives the number of terms in the digital filter.

By numerical integration of the sinh function, Figure 9.3-2, Kaiser found Table 9.3-1. The overshoot gives the attenuation, and the first crossing into this admissible band gives $\Delta F/2$ and hence D. From this table Kaiser derived the empirical formulas for α and D, α as in the preceding and D by

$$D = \begin{cases} \dfrac{A - 7.95}{14.36}, & A > 21 \\ 0.9222, & A < 21 \end{cases}$$

Solving for N, we get (9.3-7)

$$N = \frac{D}{2\Delta F} = \frac{A - 7.95}{14.36(2\Delta F)}$$

for the number of terms.

9.5 DESIGN OF A BANDPASS FILTER

We have approached the design problem as if it were a low-pass (or high-pass) filter, but the same methods apply to bandpass filters with one warning, which will be mentioned later.

Suppose, as in Figure 9.5-1, that we want to design a bandpass filter. As usual, we expand the ideal transfer function in the Fourier series (9.1-5)

$$\tilde{H}(f) = c_0 + 2 \sum_{k=1}^{N} c_k \cos 2\pi kf \tag{9.5-1}$$

where the c_k are the coefficients in the digital filter being designed. We get for the c_k

$$c_0 = 2(f_s - f_p)$$

$$c_k = \frac{1}{k\pi} [\sin 2\pi kf_s - \sin 2\pi kf_p] \tag{9.5-1}$$

Note that the limit as $k \to 0$ gives $c_k \to c_0$.

For a low-pass filter, $f_p = 0$; and for a high-pass filter, $f_s = \frac{1}{2}$.

According to Section 9.3, we compute the following in turn. Using (9.3-5), from the δ we compute A. Then from (9.3-7) we compute N, the number of terms. We round this value up. This gives a filter with $2N + 1$ terms. Next we find α from (9.3-6) to get the window shape. Then we find the window weights w_k from (9.3-1), with $w_0 = 1$, $w_N = 1/I_0(\alpha)$, and $w_{-k} = w_k$. The products of the c_k from (9.5-1) times the w_k are the final filter coefficients. Note that we do not need the $W(f)$.

Having done all this, we then check the formulas by evaluating the Fourier series of the transfer function

$$\tilde{H}(f) = c_0 + 2 \sum_{k=1}^{N} c_k w_k \cos 2\pi kf$$

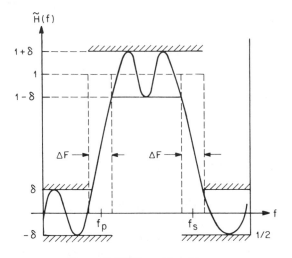

FIGURE 9.5-1 BAND PASS FILTER

We will find that occasionally the transfer function from (9.5-1) has a δ that is too large. To see why a too large δ can occur in a bandpass filter, recall that the ripples came from the tails of the transform, the sinh (or sine) formula, as they are convolved into the step of the ideal filter. Since there are two discontinuities for each band (positive and negative frequency), four sets of ripples are created by the tails while the window is being convolved into the ideal filter characteristic. In the worst case, these ripples can combine and make the total ripple at most twice as large as it should be, 6 dB larger than we designed for. Clearly, this effect depends on the spacing of the two vertical sides of the bandpass filter, the spacing between the positive and negative frequency bands, and the spacing of the ripples in the window function (the N we use). If this error cannot be tolerated, we can redesign the filter, starting with a smaller δ to allow for this effect.

9.6 REVIEW OF KAISER WINDOW FILTER DESIGN

Because of its importance, we will review all the steps in the design of filters using the Kaiser window. This review also provides the basis for a computer program (one has been written for so simple a computer as the TI-59).

First draw Figure 9.5-1, where for a low-pass filter $f_p = 0$ and for a high-pass filter $f_s = \frac{1}{2}$. Enter the four design parameters δ, ΔF, f_p, and f_s. Compute in turn:

1. $A = -20 \log_{10} \delta$

2. $N = \dfrac{A - 7.95}{28.72 \, \Delta F}$ (round up)

If N is acceptable:

3. $\alpha = \begin{cases} 0.1102(A - 8.7) & 50 \le A \\ 0.5842(A - 21)^{0.4} + 0.07886(A - 21) & 21 < A < 50 \\ 0 & A \le 21 \end{cases}$

4. $c_0 = 2(f_s - f_p)$

    $c_k = \dfrac{1}{\pi k} [\sin 2\pi k f_s - \sin 2\pi k f_p]$ $(k = 1, 2, \dots, N)$

5. $I_0(\alpha)$. Use subroutine.

6. $w_k = \begin{cases} \dfrac{I_0[\alpha \sqrt{1 - (k/N)^2}]}{I_0(\alpha)} & (|k| \le N) \\ 0 & (|k| > N) \end{cases}$

7. $\tilde{H}(f) = c_0 + 2 \displaystyle\sum_{k=1}^{N} c_k w_k \cos 2\pi k f$ $(0 \le f \le \frac{1}{2})$

    Spacing in f at least as fine as $\frac{1}{2}N$

If this curve is not satisfactory, see end of Section 9.5.
    The subroutine for $I_0(x)$:

$$I_0(x) = 1 + \sum_{n=1}^{\infty} \left[\frac{(x/2)^n}{n!} \right]^2 = 1 + \sum_{n=1}^{\infty} u_n$$

Compute $(x/2)$. Store.
Set

$$\begin{cases} n = 0 \\ S_0 = 1 \\ u_0 = 1 \end{cases}$$

Do for

$$n = 1, 2, \ldots.$$

$$n = n + 1$$

$$u_n = u_{n-1} \left[\frac{x/2}{n} \right]^2$$

$$s_n = s_{n-1} + u_n$$

Test for end of loop.

$$| u_n / s_n | \leq \epsilon \qquad (\epsilon \text{ your choice})$$

Note: The series for $I_0(x)$ converges rapidly *once* the terms start decreasing $(n > x/2)$.

9.7 THE SAME DIFFERENTIATOR AGAIN

In Section 6.4 we designed a differentiator that cut off ideally at $f = 0.2$ to remove noise in the upper 60% of the frequency band. Let us see how well the Kaiser window handles this same problem.

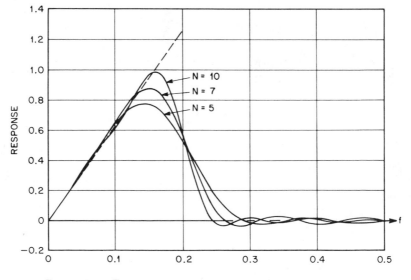

FIGURE 9.7-1 FREQUENCY RESPONSE OF A 30-dB DIFFERENTIATOR-SMOOTHING DIGITAL FILTER

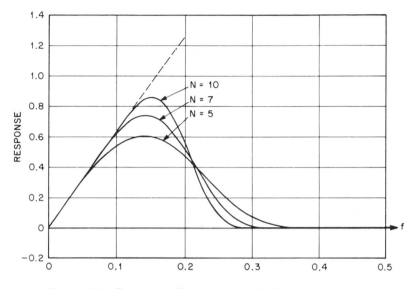

FIGURE 9.7-2 FREQUENCY RESPONSE OF A 50-dB DIFFERENTIATOR-SMOOTHING DIGITAL FILTER

For comparison purposes, we will use the number of coefficients rather than the transition bandwidth. (Solving for the transition width in terms of N is not difficult.)

Figures 9.7-1 and 9.7-2 show the transfer functions in cases of $A = 30$ dB and 50 dB for $N = 5, 7, 10$ terms. The higher flatness at the upper end of the Nyquist interval is purchased at the price of a wider transition band. This factor is an asset if the noise is confined to the upper part of the interval, but if it is spread across the whole band, matters are rather different. In Figure 9.7-3 we see another comparison of filters for $N = 5$. Finally, in Figure 9.7-4 we have another comparison.

To test the filter on some data, we will use the same noise as in Section 6.4 so that the results can be compared easily. See Figures 9.7-5 and 9.7-6.

9.8 A PARTICULAR CASE
OF DIFFERENTIATION

This example is taken from nuclear physics and is given to provide some of the flavor of a real problem in filter design. The original data consisted of counts of nuclear events classified according to the energy of the particles. Thus we are given a table of how many particles have energies in each of

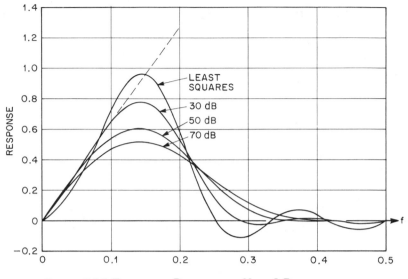

FIGURE **9.7-3** FREQUENCY RESPONSE OF $N = 5$ DIFFERENTIATOR-
SMOOTHING DIGITAL FILTER

various intervals. Furthermore, we can expect that the experiment was run for as long as practical but not as long as one could wish. Moreover, in the energy intervals (equally spaced). We see immediately that such data coming from random events must be "noisy," since from experiment to experiment we would obtain rather different counts of the events in the

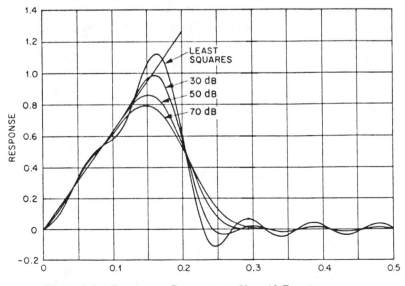

FIGURE **9.7-4** FREQUENCY RESPONSE OF $N = 10$ DIFFERENTIATOR-
SMOOTHING DIGITAL FILTER

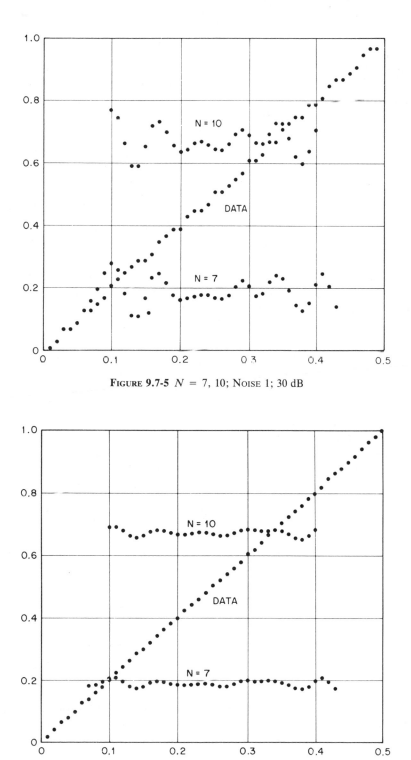

FIGURE 9.7-5 $N = 7, 10$; NOISE 1; 30 dB

FIGURE 9.7-6 $N = 7, 10$; NOISE 2; 30 dB

practice, the square roots of the counts should be used, for they will have equal variance in the various intervals. However, we will not confuse the reader with this detail.

When, in addition to this basic noise of the underlying phenomenon, we observe that the physically meaningful quantity is the derivative of the nuclear counts, we see that we have a very noisy situation that requires care in the processing of the data.

To measure where the signal and noise can be separated, two steps were taken. First, a spectrum of the raw data was computed (see Chapter 11). It showed that the spectrum was essentially flat past about the first 5% of the frequency interval. Second, a theoretical shape of the data was tried and again analyzed by the spectrum program. It, too, showed about the same breakpoint between the message and the noise. Why do we feel that a flat spectrum implies noise? Simply, if we have a noise spectrum falling off rather slowly, then the aliasing that arises from the size of the intervals (the folding back and forth of the spectrum) will usually tend to be flat across the whole spectrum, while the message in a properly designed experiment will not be aliased. (Section 11.6)

A differentiator filter was therefore designed that rose linearly from approximately $f = -\frac{1}{40}$ to $\frac{1}{40}$ in the frequency domain and was zero elsewhere. The Fourier expansion was naturally in sines with a spare factor i [which is necessary, since the derivative of $\exp(2\pi i f t)$ has a factor i]. This expansion, which varied, depending on the exact placement of the edge (so we could adjust this cutoff point), then served as the basis for the design. The Kaiser window was used to get the final filter design.

The nuclear counts were grouped into equal-sized intervals, and it was the total in each interval that was recorded. This procedure is not the same as recording samples of the function. However, if we imagine that the rectangular window is convolved across the original function and that the result is sampled, we find that we have the number of counts in the interval. Thus the grouping of the data is equivalent to a preliminary pass of a rectangular window and should be allowed for in the final interpretation.

As judged by the physicist, the results were satisfactory (of course, better data was desired, but the filter design caused no complaints). There is some risk that, by adjusting the cutoff edge on a single run of data, we will be fooled as to the accuracy; however, by applying it to several runs taken under different conditions, the effect of moving the edge was understood.

This example illustrates a number of important factors. First, the data need not be a time series but can be a series of equally spaced data in some other variable. Second, using the preceding methods of filter design, almost all the noise can be successfully eliminated from the interval where it does not overlap the signal. Thus filters apply not only to time series but to other forms of equally spaced data as well.

9.9 OPTIMIZING A DESIGN

Up to now we have adopted simple design criteria, the closeness (in a somewhat unspecified sense) of the actual transfer function to the ideal square-cornered transfer function. Both the raised cosine (with or without the platform) and the Kaiser windows are means of changing the Fourier series (least-squares) fit of the original design to reduce ripples (but, of course, they increase the error in a least-squares sense). In practice, there are numerous ways of specifying the properties desired for a particular filter.

To meet the different criteria, many design methods exist and more are easily invented. But such special methods are beyond the scope of this book. There is, fortunately, a very elementary design method that meets many such needs:

1. Decide on the kind of criterion that you want to apply and express it as a positive number with the reasonable order relationship that better designs have larger numbers.

2. Find a first approximation to the filter that you want from the Kaiser design.

3. Optimize the design by varying the adjustable parameters c_k of the filter, using (1) as a guide.

This elementary design method has a number of possible disadvantages. First, a great deal of machine time may be needed. Second, the optimization comes to a *local* maximum, and it is possible that there is a much better design that can be reached from some other starting point. Finally, you have little or no idea of what similar filters with a slightly different number of terms will do unless you repeat the whole process; thus you are blindly searching for an optimum with almost no theory to guide you. This lack of elegance makes the experts tend to avoid the method; but for the beginner and the occasional designer it has the great advantage of wide applicability and, in addition, it requires comparatively little special knowledge. Computers are cheap these days.

In real life the difficulties are generally the choice of the criteria and the first trial transfer function to use and not the amount of machine time. Kaiser's design method should help guide the beginner in an actual situation toward making the choice of the starting filter.

The slowness of the process of finding the optimum is less important. A maximum that is difficult to locate suggests a "delicate" design. But note: as Figure 9.9-1 shows clearly "mathematically delicate" means "practical robustness"!

However, the lack of surrounding theory and the risk of finding a local

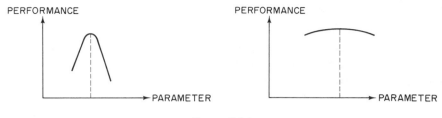

FIGURE 9.9-1

optimum that is far from the best are serious matters that should be considered in each application.

9.10   A CRUDE METHOD OF OPTIMIZING

The following is how *not* to optimize. It is here to warn you against trying the obvious methods that experience has repeatedly shown are not adequate in many cases.

Suppose that we are given both an evaluation function E that measures how well the criterion is met (E gives nonnegative values and increasing E means a better design) and a first approximation to the design with the N parameters $p_1, p_2, \ldots, p_N$ (usually the coefficients c_k of the filter).

We set up three arrays in a computer: (1) the array of N initially guessed coefficients (parameters) p_k from the approximate design; (2) a corresponding array of N estimates of how much we think the coefficients might be in error Δp_k (these guesses will rapidly be eliminated and replaced with more accurate guesses generated by the optimization routine so that bad guesses merely lead to extra computations); and (3) an array of the N improvements due to a change in the parameters (the third array is later filled in).

The first step in the computation is to evaluate the function E for the given coefficients p_k. We put $1/N$ of this amount into each place in the third array, where N is the number of parameters p_k.

The next step is to change the first parameter p_1 by the amount Δp_1, which is the entry in the uncertainty array corresponding to the first parameter, and again evaluate the function E. If the difference in the two E values is positive, we have found an improvement. Therefore, we take another step of the same size in the same direction along the p_1 axis. We continue until we find an E value that is $\leq$ the previous one (note the $\leq$). We then have three consecutive values of the parameter p_1, which we will label p_1^-, p_1^0, p_1^+. If the first change in E was a decrease or the same value, we interchange the starting and first trial values, along with the sign of Δp_1, and continue going in this direction, searching as before for a relative maximum of E as a function of p_1. When we find it, we are in the same position as before. If we discover that the value of E falls when we move in either direction, we have a situation similar to the preceding situation of three values with the larger value in the middle.

Through the three consecutive values we pass a parabola

$$E(p_1) = A + B(p_1 - p_1^0) + C(p - p_1^0)^2$$

where p_1^0 is the middle value of the three values of the coordinate of the parameter p_1. The maximum of the parabola occurs at the value where

$$\frac{dE(p_1)}{dp_1} = 0 = B + 2C(p_1 - p_1^0)$$

Thus the maximum occurs at

$$p_1 = p_1^0 - \frac{B}{2C}$$

From the three given values it is easy to see that the coefficients of the parabola are given by

$$B = \frac{E(p_1^+) - E(p_1^-)}{2d}$$

$$C = \frac{E(p_1^+) - 2E(p_1^0) + E(p_0^-)}{2d^2}$$

where (the step spacing)

$$d = p_1^0 - p_1^- = p_1^+ - p_1^0$$

From our method of search we avoided having $C = 0$, and so we are not dividing by 0 (unless all three values of E are the same!). We now have a new p_1.

Having temporarily optimized the first parameter, we need to refill the corresponding values in the three arrays. Obviously, the maximum of the parabola is taken as the new entry p_1 in the first array. We could pick one-half the change from the initial to the final estimates of p_1 as the new uncertainty, Δp_1, but we must avoid having the new uncertainty be zero, for if it were, no further search would occur. Thus we take the larger of one-half of the change and one-half of the initial uncertainty as our new uncertainty. Finally, we select the change in the function E from its initial value to the value at the new point as the corresponding entry in the third array. Again we need to worry about a zero entry, since, as we shall see, it would be impossible to resume examining improvements in the parameter if it were zero. To avoid this trouble, we again adopt a selection of the larger of one-half of the original change in E and the current amount of the improvement in E.

With the first parameter processed, we repeat the procedure for the second, the third, and so on, through the entire set of parameters. We end up with three new arrays: the first has the new estimates for the parameters; the second has the uncertainties that reflect how much we moved from our initial guess to the new one; and the third array has an indication of where we achieved large and where we achieved small improvements. So once we have gone through the array, we have available the computer-produced estimates of the parameter values, the uncertainties, and where to expect the most improvement.

From this point on we scan the improvement array for the largest value and work on the corresponding parameter. As a result, at every stage of the computation we tend to have about the same order of magnitude in all the improvements in the third array. The corresponding step sizes that produce the improvements indicate the sensitivity of the fit in the parameters. When the improvements become small per computation cycle, we cannot expect that E will get more than the maximum allowed for the model that we are testing, and we stop further computation. Gradually increasing the values in the improvement column is worthwhile in order that a chance small value will not prevent us from looking there later.

Notice that this algorithm has a number of invariances. It rapidly eliminates the starting values, both the positions and the uncertainties, and also omits the chance order in which we selected the parameters of the problem. At the end of the computation each new parameter change would have about the same amount of effect in improving the fit as any other parameter.

This simple optimizer illustrates several important points. First, it is the kind of optimizing process that the beginner is apt to believe is good. And it is good for simple problems.

Second, it shows how a simple optimizer can fail. Suppose, for instance (as we will be doing in Chapter 12), that we are trying to reduce the maximum ripple of a polynomial in some interval. It is probable that one parameter, say p_1, will tend to reduce one peak (or valley) of the polynomial while increasing the size of one or more other extremes. Thus changes in p_1 will work *until* some second extreme comes up to the same value as the first. There will be a sharp "corner" to the surface at this point. It is also likely that some other parameter, say p_2, will tend to reduce the second extreme while at the same time increasing the first. Thus changes in neither parameter can by themselves improve matters once the two extremes are of the same size: only a combination of simultaneous changes in both will further reduce the maximum extreme of the polynomial in the given interval. And, of course, the more equal-sized extremes there are, the more parameters must simultaneously be changed if further improvements are to be found. The simple, one-parameter-at-a-time optimizer is clearly inadequate.

The third point is that powerful optimizers in the local library tend to be "powerful" in one sense or more. They may try to reach the optimum rapidly, but this fact only saves machine time and is not apt to be important to the occasional digital filter designer. They may be powerful in the sense that they will find the maximum for a wide range of surfaces, including situations where the valley being searched is both steep and curved, so that continually different directions of search must be used. In addition, the power may refer to the ability to pinpoint the maximum. This feature is often illusory. If the maximum is sharp and easy to locate, see Figure 9.9-1, the performance of the filter will change greatly for small changes in the parameter values (perhaps a roundoff to a shorter length number), whereas if it is a broad maximum, then, even though the place where the true maximum occurs is not well determined, this fact is unimportant, since almost any nearby point will give almost equally good results.

Finally, the example selected shows that the maximum surface being searched can have "corners" for changes in many of the parameters. Therefore, if as is probable, the local library optimizing routine is based on a smooth surface and fits the surface with quadratic terms in all the parameters, it is possible that the optimizer will behave peculiarly.

10

The Finite Fourier Series

10.1 INTRODUCTION

We have at times discussed the continuous functions of frequency as if it were possible to handle continuous functions in a computer, but, of course, we are usually forced to use samples of such functions. Fortunately, the sines and cosines are orthogonal over *both* the continuous interval and sets of equally spaced points. Thus we can perform analogous operations in the discrete versions of the functions used in a computer, and aliasing must be considered once more.

The problems faced in this chapter are: (1) to prove the orthogonality of the sines and cosines over a set of equally spaced points, (2) to relate the continuous and discrete expansions (via aliasing), (3) to give a sketch of the ideas behind the fast Fourier transform method of computing the coefficients of the discrete expansion (most installations have their favorite versions in the library), and (4) to give extensive warnings on its misuse. The fast Fourier transform is simply a means of computing the finite Fourier transform in essentially $N \log N$ arithmetic operations instead of (what might at first seem necessary) N^2 operations. This difference is of fundamental importance to many applications, since for large N it can mean a reduction by factors of hundreds, or thousands, or even more in machine time used, as well as a significant reduction in roundoff errors in the results.

10.2 ORTHOGONALITY

The discussion will be restricted to an even number of points, with the corresponding case of an odd number of points left to the readers in case they ever face this likely situation; there are only trivial differences between

208

the two cases. Let the $2N$ points we have be (L = length of interval)

$$x = 0, \frac{L}{2N}, \frac{2L}{2N}, \ldots, \frac{(2N-1)L}{2N}$$

Using p as the index, these numbers can be written

$$x_p = \frac{Lp}{2N} \quad (p = 0, 1, 2, \ldots, 2N-1) \qquad (10.2\text{-}1)$$

It is easiest to begin with the orthogonality of the complex exponentials. To prove that they are orthogonal, we begin with the very simple geometric progression [using the $2N$ points $x_p = Lp/2N$ (10.2-1)

$$\sum_{p=0}^{2N-1} e^{(2\pi i q/L)x_p} = \sum_{p=0}^{2N-1} e^{\pi i p q/N} \quad \text{(for integer } q) \qquad (10.2\text{-}2)$$

This geometric progression has the ratio

$$r = e^{\pi i q/N}, \qquad r^{2N} = e^{2\pi i q} = 1$$

and the sum of (10.2-2) is

$$\begin{cases} \dfrac{1 - r^{2N}}{1 - r} = 0, & r \neq 1 \\ 2N, & r = 1 \end{cases} \qquad (10.2\text{-}3)$$

The situation of $r = 1$ arises only when $q = 0, \pm 2N, \pm 4N, \ldots$. These values of q, as we shall show in Section 10.3, lead to the aliasing that we know must occur as a result of the equally spaced sampling of the function. From this simple summation (10.2-2) we move to the set of functions (k is an integer)

$$e^{(2\pi i/L)kx_p} = e^{\pi i k p/N}$$

and prove that they are orthogonal. "Orthogonal" means that the summation over the set of sample points of the kth function of the set times the *complex conjugate* of the mth is zero, except when the same function, or an equivalent aliased function, is used for both. Thus we need to prove

$$\sum_{p=0}^{2N-1} e^{(2\pi i/L)kx_p} e^{-(2\pi i/L)mx_p} = \begin{cases} 0, & |k - m| \neq 0, 2N, 4N, \ldots \\ 2N, & |k - m| = 0, 2N, 4N, \ldots \end{cases}$$

$$(10.2\text{-}4)$$

which follows immediately by writing the product of the exponentials as a single exponential and applying the previous result (10.2-3), with $q = k - m$.

To get to the "real functions" of sine and cosine, we use the Euler identities

$$\cos x = \frac{e^{ix} + e^{-ix}}{2}$$

$$\sin x = \frac{e^{ix} - e^{-ix}}{2i}$$

Writing out the sum of the product of two trigonometric functions, we obtain both the sum and difference of the exponents if the two functions are both sine or both cosine. We have for the cosines (notice the aliasing)

$$\sum_{p=0}^{2N-1} \cos\frac{\pi kp}{N} \cos\frac{\pi mp}{N} = \begin{cases} 0, & |k \pm m| \neq 0, 2N, 4N, \ldots \\ N, & |k + m| \text{ or } |k - m| = 0, 2N, 4N, \ldots \\ 2N, & |k \pm m| \text{ both } = 0, 2N, 4N \ldots \end{cases}$$

(10.2-5)

Similarly, for the sines we have

$$\sum_{p=0}^{2N-1} \sin\frac{\pi kp}{N} \sin\frac{\pi mp}{N} = \begin{cases} 0, & |k \pm m| \neq 0, 2N, 4N, \ldots \\ N, & |k - m| = 0, 2N, 4N, \ldots \\ -N, & |k + m| = 0, 2N, 4N, \ldots \\ 0, & |k \pm m| \text{ both } = 0, 2N, 4N, \ldots \end{cases}$$

(10.2-6)

The case of a sine times a cosine is easily shown to be always zero.

For the restricted set of functions

$$1, \cos\frac{2\pi x}{L}, \cos\frac{4\pi x}{L}, \ldots, \cos\frac{2\pi(N-1)x}{L}, \cos\frac{2\pi Nx}{L}$$

$$\sin\frac{2\pi x}{L}, \sin\frac{4\pi x}{L}, \ldots, \sin\frac{2\pi(N-1)x}{L}$$

(10.2-7)

only one of the conditions (10.2-5) and (10.2-6) for a nonzero sum can apply at a time, and we have the orthogonality of the $2N$ Fourier functions over the $2N$ discrete equally spaced points x_p (10.2-1). We could not reasonably expect more linear independent functions. This orthogonality is the same as we had over the continuous interval, with, of course, different normalizing factors. Thus we are led to the Fourier expansion of an arbitrary function $G(x)$ defined over the discrete set of $2N$ points x_p:

$$G(x) = \frac{A_0}{2} + \sum_{k=1}^{N-1} \left[A_k \cos \frac{2\pi kx}{L} + B_k \sin \frac{2\pi kx}{L} \right] + \frac{A_N}{2} \cos \frac{2\pi Nx}{L}$$

$$(10.2\text{-}8)$$

where we are using *capital letters for the coefficients in the discrete expansion*. Note that both the first and last cosine terms have an extra $\frac{1}{2}$ factor. We have from the orthogonality

$$A_k = \frac{1}{N} \sum_{p=0}^{2N-1} G(x_p) \cos \frac{2\pi kx_p}{L} \qquad (k = 0, 1, \ldots, N)$$

$$B_k = \frac{1}{N} \sum_{p=0}^{2N-1} G(x_p) \sin \frac{2\pi kx_p}{L} \qquad (k = 1, 2, \ldots, N-1)$$

$$(10.2\text{-}9)$$

In the complex notation we have, corresponding to (10.2-8) and (10.2-9), the equations

$$G(x) = \sum_{k=-N+1}^{N} c_k e^{2\pi ikx/L}$$

$$c_k = \frac{1}{2N} \sum_{p=0}^{2N-1} G(x_p) e^{-2\pi ix_p/L}$$

$$(10.2\text{-}10)$$

Note the great similarity of the formulas for $G(x)$ and the c_k; this corresponds to that of the Fourier integrals.

Frequently, the function $G(x)$ is given at $2N + 1$ points, *including both ends of the interval*, and we still have (because of periodicity) only $2N$ intervals. In such cases, since we have assumed that the function is periodic, we usually take the true end value as the average of the two endpoints,

$$\frac{G(0) + G(L)}{2}$$

Of course, if the function is periodic, then both end values are the same, and this is simply one of the values. The effect of this averaging is to expand slightly the basic formulas for the coefficients of A_k (10.2-9) without changing the B_k.

We see that the summation formula for A_0 in the discrete case corresponds to the trapezoid rule of integration. Remember that the periodicity covers the two end values of weight $\frac{1}{2}$.

If we used the midpoint formula rather than the trapezoid rule for approximating the integral, the preceding arguments would lead to a corresponding set of orthogonality relations over these midpoints of the intervals. For these points the highest frequency cosine term will be identically zero, and we need instead to include the corresponding sine term in the basic set of functions.

Exercises

10.2-1   Sum the geometric progression over $2N + 1$ points.

10.2-2   Prove the orthogonality of the sines and cosines over $2N + 1$ points.

10.2-3   Expand

$$G(x) = \begin{cases} -1, & -N < x \le 0 \\ 1, & 0 < x \le N \end{cases}$$

10.2-4   Expand

$$G(x) = x, \qquad -N < x \le N$$

10.2-5   Expand

$$G(x) = \begin{cases} 1, & |k| < N/2 \\ \frac{1}{2}, & |k| = N/2 \\ 0, & |k| > N/2 \end{cases}$$

where N is an even number.

10.3   RELATIONSHIP BETWEEN THE DISCRETE AND CONTINUOUS EXPANSIONS

Since it is occasionally necessary to replace the continuous functions of our theory with the discrete functions of computing practice, we investigate their mutual relationship. Suppose that the continuous expansion of $G(x)$ is given by

$$G(x) = \frac{a_0}{2} + \sum_{k=1}^{\infty} \left[a_k \cos \frac{2\pi kx}{L} + b_k \sin \frac{2\pi kx}{L} \right] \qquad (10.3\text{-}1)$$

where we have used lowercase letters for the coefficients of the continuous function expansion and will continue to use uppercase letters for the discrete expansion. If we multiply both sides of this equation by $\cos 2\pi kx_p/L$ and sum, we obtain, due to the aliasing given in the previous section (10.2-5),

$$\sum_{p=0}^{2N-1} G(x_p) \cos\left(\frac{2\pi k x_p}{L}\right) = NA_k - N(a_k + a_{2N-k} + a_{2N+k} + \cdots)$$

from which we see that the coefficient calculated in the expansion A_k is given by

$$A_k = a_k + \sum_{m=1}^{\infty} (a_{2Nm-k} + a_{2Nm+k}) \tag{10.3-2}$$

For each k this formula gives the calculated result A_k in terms of the true answers a_k. It shows the usual aliasing in a slightly new form and offers a different derivation of the aliasing effect. The aliasing is illustrated in Figure 10.3-1. Imagine the frequencies as being on a ribbon to the right (and of course to the left). The ribbon is then folded into a bin as shown. Frequency values that fall on top of each other appear, due to aliasing, as the same frequency.

Similar calculations using the sine (10.2-6) in place of the cosine lead to the formula

$$B_k = b_k + \sum_{m=1}^{\infty} (-b_{2Nm-k} + b_{2Nm+k}) \tag{10.3-3}$$

The constant term gives an especially important result, for it shows the relationship in the frequency domain between the integral of a periodic function and its trapezoid rule approximate calculation value A_0:

$$A_0 = a_0 + 2 \sum_{m=1}^{\infty} a_{2Nm} \tag{10.3-4}$$

If the higher-indexed coefficients are small, then the error is small.

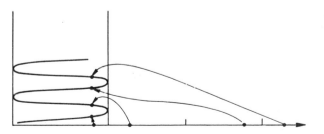

FIGURE 10.3-1 ALIASING

Thus we see the role played by the Nyquist folding frequency in a different light; we can act as if we had a continuous function, but the process of sampling will bring us into the Nyquist interval (or any other equivalent interval that we care to choose).

Exercise

10.3-1 Carry out all the details for the aliasing expressions for A_k and B_k.

10.4 THE FAST FOURIER TRANSFORM

The direct approach to the calculation of the Fourier coefficients of a discrete expansion appears at first (10.2-9) to be of the order N^2 operations; there are $2N$ coefficients and $2N$ terms in each summation. A recent rediscovery and adequate presentation by Cooley and Tukey of a method for calculating the Fourier coefficients reduces the amount of computing to the order of $N \log N$. This amount can be significant in a large problem, often well over a factor of 100 in machine time saved, thus well less than 1% of the original time.

Research is still being done on how to extract the last bit of savings of machine time, but this factor is of little interest in an introductory course. We therefore present the main idea and avoid becoming involved in details. As a rule, the local computer library has a fast Fourier transform program. It is important to emphasize, however, that within roundoff the results of the straightforward method and the fast Fourier transform method *are exactly the same; they are simply alternate ways of computing the same thing.* The fast Fourier transform, because it does less arithmetic, typically has less roundoff error in the results. For convenience, we will set $L = 1$ so that $x_p = p/2N$.

We assume that the number of sample points of the function, $2N$ (or $2N + 1$), can be factored into two integers greater than 1:

$$2N = PQ$$

Thus the sample points are

$$x_m = \frac{m}{PQ}$$

The Fourier coefficients of the expansion of the function $G(x)$ are

$$A_k = A(k) = \frac{1}{PQ} \sum_{m=0}^{PQ-1} G(x_m) e^{-2\pi i k x_m}$$

We now do the decisive step and divide k by P in order to obtain the quotient k_1 and the remainder k_0. The next step is to divide m by Q and get the quotient m_1 and the remainder m_0. Both k and m are *uniquely* represented when we write the two summation variables as

$$k = k_0 + k_1 P$$

$$m = m_0 + m_1 Q$$

where k_1 is the quotient on dividing k by P and k_0 is the remainder. Similarly for m on division by Q. We have the conditions, of course,

$$k_0 < P \quad \text{and} \quad k_1 < Q$$

$$m_0 < Q \quad \text{and} \quad m_1 < P$$

Using these representations for k and m, we then have for the summation (note that we have taken into account all the terms but have used the essential relation $e^{-2\pi i k_1 m_1} = 1$)

$$A(k_0 + k_1 P)$$

$$= \frac{1}{PQ} \sum_{m_0=0}^{Q-1} \left[\sum_{m_1=0}^{P-1} G\left(\frac{m_0 + m_1 Q}{PQ}\right) e^{-2\pi i(k_0 + k_1 P)(m_0 + m_1 Q)/PQ} \right]$$

$$= \frac{1}{PQ} \left[\sum_{m_0=0}^{Q-1} e^{-2\pi i k_0 m_0/PQ} e^{-2\pi i k_1 m_0/Q} \right] \left[\sum_{m_1=0}^{P-1} G\left(\frac{m_0}{PQ} + \frac{m_1}{P}\right) e^{-2\pi i k_0 m_1/P} \right]$$

It is easy to recognize that the quantity in the second set of brackets is a Fourier expansion involving $1/Q$ of the samples of the function, phase shifted m_0/PQ. Furthermore, since $0 \leq m_0 < Q$, there are exactly Q such sums to be done. By labeling these Q sums as

$$\overline{A}(k_0, m_0)$$

we will have the formula

$$A(k_0 + k_1 P) = \frac{1}{PQ} \sum_{m_0=0}^{Q-1} \overline{A}(k_0, m_0) e^{-2\pi i[k_0/PQ) + (k_1/Q)]m_0}$$

This is a second Fourier series expansion computation, and this time, since $0 \leq k_0 < P$, there are P such sums to be evaluated.

A count of the number of arithmetic operations shows that it is proportional to

$$PQ(P + Q)$$

This amount of computation is in place of the original amount, which was proportional to $(PQ)^2$.

Evidently, for each factor P_i greater than 1 that we can find in the number $2N$, we can repeat this process. In the end we will arrive at, approximately,

$$P_1 P_2 P_3 \cdots P_k (P_1 + P_2 + \cdots + P_k)$$

arithmetic operations instead of the original $(2N)^2$ operations. Thus we see that when the number of sample points $2N$ is a power of 2, we will obtain the maximum reduction of computation, and the number of terms appearing in the sum inside the parentheses is the log $(2N)$.

Experience shows that this simple trick covers most of the savings (in a multiplicative sense) and that additional refinements in the details will save relatively less. However, if the Fourier transform is to be done frequently, further tricks can and should be used to reduce roundoff even if the computing time saved is not important. See references [Br], [IEEE-1], and [IEEE-2].

The fast Fourier transform provides a popular method of filter design. Given the original data x_n, we take the Fourier transform of the x_n. Next, we take the values, which are the Fourier expansion coefficients, and multiply them by their corresponding transfer function values. Since the transfer function is the curve of gain (or attenuation) of the various frequencies, this multiplication is equivalent to filtering. Then we need only apply the fast Fourier transform to these products to obtain the filtered data.

The entire process is as follows: (1) transform the data, (2) multiply the results by the transfer function, and (3) transform the products back! It is a very general design method for nonrecursive filters. The reader may, of course, be a little concerned over what happens to frequencies between the sample points of the transfer function that were used. Yet if the points are densely placed, there should be no problem (except near sudden jumps where Gibbslike wiggles can be expected). At a discontinuity of the transfer function, of course, the average of the two limits is used.

10.5 COSINE EXPANSIONS

Since the transfer functions are often even functions, that is,

$$\tilde{H}(f) = \tilde{H}(-f) \qquad (\, |f| \leq \tfrac{1}{2})$$

it is worth a special effort to look at the corresponding Fourier expansions.

All the coefficients of the sine terms are zero. This result is immediately obvious from the fact that the sine is an odd function of its argument and that the product of an even and an odd function is an odd function, so that the integration over a symmetric interval is zero. Thus only the cosine terms need be considered. Again, from the evenness, this time we see that the summations need be done over only half the range, and we will obtain, corresponding to (10.2-9),

$$A_k = \frac{1}{2N}\left[G(0) + 2 \sum_{p=1}^{N-1} G\!\left(\frac{p}{2N}\right) e^{-2\pi ipk/2N} + G\!\left(\frac{1}{2}\right) \right]$$

The corresponding formula for the midpoint integration method is

$$A_k = \frac{1}{N} \sum_{p=0}^{N-1} G\!\left(\frac{p + 1/2}{2N}\right) e^{-(2\pi i/2N)(p + 1/2)k}$$

We shall see these formulas in a different situation in the next chapter.

Exercise

10.5-1 Work out the formulas of Section 10.5 for the odd transfer function $\tilde{H}(f) = -\tilde{H}(-f)$.

10.6 ANOTHER METHOD OF DESIGN

You can design a filter another way. Just pick the equally spaced samples from the transfer function and perform the fast Fourier transform. You have the coefficients of the filter, since they are the coefficients of the Fourier expansion of the transfer function (with the small adjustments we have always been making). In this case you can find out what happens to all the frequencies you want by plotting the transfer function at the corresponding values you are interested in.

10.7 PADDING OUT ZEROS

The usual FFT program in the library assumes that you have exactly 2^n samples of the input data. Often you do not have exactly that amount and you face the choice of either discarding some of the gathered data or else supplying the needed values. The usual, and unfortunately widely used, method is to supply the missing data in the form of zeros; indeed the program may automatically do this for you!

But think! This function with its supplied zeros does not resemble the true data—the Fourier coefficients that you compute will not be near to those of the correct function. As a trivial illustration of this fact, consider an input function that is identically 1, but you had to supply a run of missing data and made them all 0. You are expanding a discontinuous function, and you know intuitively that before aliasing the computed coefficients will fall off like $1/k$. The details of the computation will, of course, depend on where you start the run of missing zeros. But looking at formulas (10.2-9) you see that you should get 0 for all coefficients except the A_0; but due to the missing data you will get the negative of the sum of these corresponding trigonometric functions. Instead of getting just the zero frequency, you will almost certainly get all the coefficients as not being zero, and often by a considerable amount!

A similar situation occurs for a pure sine or cosine as an input. Ideally you should get a single nonzero output, but in fact you will probably get all the coefficients not zero. Again, you can expect the coefficients to fall off like $1/k$, but due to the finite range and aliasing, this will be masked a great deal.

Padding with zeros to cover the missing data will not give you reasonable results. Illustrating this point is not only a simple mental exercise, you can easily verify it experimentally. Each pure frequency in your input signal, due to the linearity of the Fourier series, will behave as indicated above, and collectively they may give you almost anything. Take a run of your data that covers 2^n points and analyze it by the FFT. Then remove a reasonable fraction at one end and replace this by zeros and analyze again. It is likely, though not certain, that the results will differ considerably!

It is perhaps necessary to remind you that the finite Fourier series assumes that your input function is periodic with the period exactly matching the interval. But the interval was determined by the product of the sampling interval and the number of intervals used—and this may have very little to do with your signal and its frequencies! Furthermore, you had to sample at a rate sufficiently high so that serious aliasing does not occur. We say again, the FFT is simply a different, efficient method of computing the finite Fourier series—it has no magical properties.

Exercises

10.7-1 Consider padding zeros at both ends. Hint: Consider Lanczos' rectangular window.

10.7-2 Use 10.7-1 to analyze padding $g(x) = \sin kx$ both continuous and discrete.

11

The Spectrum

11.1 REVIEW

As we remarked in the Preface, the spectrum has been widely and successfully used to look inside "black boxes," both those of Nature and manmade ones. It is perhaps the single most useful tool for this purpose.

Given a signal $g(t)$, say of time, but not necessarily so, we have the Fourier integral representation (see Chapter 9)

$$g(t) = \int_{-\infty}^{\infty} G(f)e^{2\pi ift} \, dt$$

where $G(f)$ is the amount of frequency f that is in the signal $g(t)$. The quantity

$$| G(f) |^2$$

is often called the *power* in the signal at frequency f, and the whole curve, as a function of f, is called the *power spectrum*. Most of the time it is the square of the absolute value that is plotted as the spectrum, but occasionally you will see just the absolute value called the power; it is necessary to be careful when you read different books or articles to note which the author is using.

For the Fourier series we saw (Section 2.4) that it was not the individual coefficients a_k and b_k that mattered, but rather that

$$a_k^2 + b_k^2$$

was the invariant with respect to the choice of the origin of the coordinate system. This corresponds to the power spectrum of the Fourier integral, but of course is defined only for integer k and is not a continuous function of k. This difference arises from the assumption that the Fourier series represents a periodic function, whereas the Fourier integral can represent any function, periodic or not.

Thus, although the spectrum is only a related topic to digital filters, it is necessary to discuss it carefully. This is especially true because there are many misunderstandings and misuses of power spectra. In a sense this chapter is devoted to warning you about the misuse of spectra because of the failure to understand just what is computed. We, therefore, take up one by one the various effects that arise in the computation of the spectrum of a function. Many of the necessary mathematical tools are now available for this task.

11.2   FINITE SAMPLE EFFECTS

The first thing to note is that in practice you have almost always only a finite sample of the function you are studying. Section 8.8 showed that if the sample runs from $-T$ to $+T$ this is equivalent to multiplying the original infinitely long time function by a rectangular window of this length and, therefore, by the convolution theorem (Section 8.7), convolving (smearing) the spectrum by the function (window)

$$\frac{(2T) \sin[2\pi T(f_1 - f)]}{2\pi T(f_1 - f)}$$

where f_1 is the variable to be convolved with the original $G(f_1)$.

The larger T (half the length of the observation) is, the narrower is the main lobe of the convolving window, and the more the *resolving power* of the window. It is directly proportional to the length of run of data. Of course, the theory of windows suggests that at times a shape other than the simple rectangular shape might be preferable. Usually you can gain in having smaller side lobes by permitting a wider main lobe.

Although you want to look at the spectrum locally (the main lobe), the side lobes of the window permit *leakage* from more remote parts of the spectrum. This again raises the question of shaping the initial window, which we cannot go into here; it is fundamentally a statistical question and lies outside a first course in digital filters. We merely note that the von Hann window greatly reduces the size of the side lobes at the cost of a main lobe of double width. The Hamming window is used mainly when there is a very strong line in the spectrum and you do not want it leaking through and affecting what you are looking at locally. The price of this is, however, a great deal of leakage, on the average, since the tails of the Hamming window are very much larger than those of the von Hann window.

There is a vast literature on the topic of windows and many *figures of merit* for comparing windows. References can be found in [IEEE-1] and [IEEE-2] and we will not discuss the matter further in this book.

11.3 ALIASING

In Section 2.2 we carefully gave both a mathematical and a physical description of the important effect of aliasing due to the sampling at equally spaced points of the original signal. In Section 10.3 we gave formulas showing exactly the aliasing of the Fourier coefficients when you pass from the continuous to the sampled data. If you think of the continuous signal being first limited in range and then sampled, you see that the effect on the convolving window of the previous section is to replace it with (see equation 5.5-6) the window

$$\frac{\sin(N + 1/2)\theta}{[(2N + 1) \sin(\theta/2)]}$$

where we have normalized the window to have unit area to make the expression comparable.

The sampling can be thought of as occurring before or after the limiting of the length of the data. If before, we are limiting the already aliased frequencies; if after, we are aliasing the corresponding window of Section 11.2. In either case the result is the same; the preceding is the effective window. The relationship depends on the formula

$$\frac{z}{\sin z} = 1 + z \sum_1^\infty (-1)^n \left\{ \frac{1}{z - \pi n} + \frac{1}{z + \pi n} \right\}$$

11.4 COMPUTING THE SPECTRUM

The spectrum is generally computed from the finite, sampled data via the fast Fourier transform (Section 10.4). It is necessary to say again that this is mathematically exactly the same as finding the finite Fourier series coefficients of the function, but that numerically the fast Fourier transform generally has less roundoff error in it than has the direct computation of the coefficients. From the finite number of data points, you get the same number of Fourier coefficients, and these are exactly equivalent since from the coefficients you can reconstruct the data (within roundoff, of course). You cannot get more independent pieces of information than you put into the computation. Any interpolation you may do to construct a continuous spectrum from the isolated fast Fourier transform points gives results that must depend on the interpolation formula.

Unfortunately, the fast Fourier transform routines that are in libraries usually are designed for processing 2^n data points, and usually you have only those data points you could get and there are not a power of 2 samples. Even more unfortunately you may read in the description of the routine that, if you do not have the right number of data points, you should fill in the rest (or it is automatically done for you by the program) with zero values. This is almost always the same as putting sharp discontinuities into the function.

There is a simple way of seeing this effect theoretically. First, you can think of it as the complete 2^n data points multiplied by a rectangular window that goes to zero where your actual run of data ends. Thus, by the convolving theorem, this is equivalent to convolving the true finite Fourier coefficients by those of the rectangular window, and we have seen (Section 4.6) that this window produces a falling off of the Fourier coefficients like $1/k$.

Even if you have exactly a power of 2 data points, remember that the finite Fourier series supposes that the function is periodic and is wrapped around a cylinder, and that there may be a discontinuity where the start and end of the data meet; see Section 11.5.

This brings up the obvious question of suitably windowing the data to remove as much of the effect of the discontinuity as you can. It *seems* to the author (published advice notwithstanding) that the best you can do is "fair in" a suitably raised cosine between the two ends of the data. The formula to use is

$$a + b \cos \frac{\pi k}{M}$$

where k runs from 0 to M. The a and b are to be chosen to make the data continuous at the ends of the interval of length M ($M - 1$ points are being faired in). In any case, if your data is reasonably well removed from zero, it is foolish to fair in at both ends a run of zeros. If the interval to be filled in is very narrow (conventional wisdom suggests at least 20% of the total run length), you may want to make the faired-in cosine cover part of the gathered data. If you want to be very careful and if you have a strong trend in the data, then, of course, you should make a greater effort to match slopes where you are inserting data and doing the fairing; the purpose is to end up with a function that does not have sharp discontinuities in either the function or the first derivative.

What is required is simple, *clear thinking on your part*. What is the actual function you are using to compute the finite Fourier series? Not the function you want, unfortunately, but the one that the program sees is the function the fast Fourier transform works on. And remember that you are getting the finite number of coefficients from the finite number of data points supplied; you are *not* getting and *cannot* get a continuous function out of the computation.

11.5 NONHARMONIC FREQUENCIES

The finite Fourier series *assumes* that the function is periodic. It computes the amount of each harmonic frequency present in this periodic function. The spectrum you get,

$$| c_k c_{-k} |$$

or in real terms

$$a_k^2 + b_k^2 \quad \text{or} \quad A_k^2 + B_k^2$$

consists of the discrete values for each k. And the range of k is finite; it is not infinite. Essentially, you get out half the number of input data points. Beware of thinking that the coefficients you get from the fast Fourier transform computation are giving you only information about the signal. They are also giving you information about: (1) the finite length of data, (2) how you supplied missing data values, and (3) your *assumption* of periodicity!

What happens when the original signal contains a nonharmonic component (nonharmonic with respect to the interval actually chosen for the computation)? We will examine the mathematically simple case of the continuous Fourier series. Let the nonharmonic signal be

$$\cos(\omega t + \phi)$$

The complex Fourier coefficient c_k is given by the formula

$$
\begin{aligned}
c_k &= \frac{1}{2\pi} \int_{-\pi}^{\pi} \cos(\omega t + \phi) e^{-ikt}\, dt \\
&= \frac{1}{4\pi} \int_{-\pi}^{\pi} [e^{i\phi} e^{i(\omega - k)t} + e^{-i\phi} e^{-i(\omega + k)t}]\, dt \\
&= \frac{i}{2\pi} \left[e^{i\phi} \frac{\sin(\omega - k)\pi}{\omega - k} + e^{-i\phi} \frac{\sin(\omega + k)\pi}{\omega + k} \right]
\end{aligned}
$$

Using the fact that $\sin \pi k = 0$ and $\cos \pi k = (-1)^k$, we can expand the trigonometric functions and simplify as follows.

$$c_k = \frac{i}{2\pi} \sin \omega\pi (-1)^k \left[\frac{e^{i\phi}}{\omega - k} + \frac{e^{-i\phi}}{\omega + k} \right]$$

$$= \frac{(-1)^k}{\pi} \sin \omega\pi \left[\frac{i\omega \cos \phi - k \sin \phi}{\omega^2 - k^2} \right]$$

We see immediately that there is a tendency to have the nonharmonic frequency affect most strongly the coefficients with $|k|$ near ω, and that the rate of falloff is only inversely like the distance of the integer k from the nonharmonic frequency.

To get a clearer picture, we take the special case $\phi = -\pi/2$. This makes the original function into a sine (and this starts at the zero value when $t = 0$). We then get the simple formula as a function of k

$$c_k = \frac{(-1)^k}{\pi} \left[\frac{k}{\omega^2 - k^2} \right] \sin \pi\omega$$

from which we can see how a single nonharmonic frequency appears strongly in all nearby coefficients and moderately strongly even in remote coefficients. To see how this mathematical expression resembles your intuition, you have only to think of the picture of the integrand being computed to get the coefficients and note how the nonharmonic function is *not* orthogonal to the corresponding Fourier function being used to find the corresponding coefficient.

Exercises

11.5-1 Set $\phi = 0$ and analyze the results.

11.5-2 Analyze for the sum of the two frequencies.

11.6 REMOVAL OF THE MEAN

Given some data, most statisticians immediately remove the mean from the data. It may also be true that the way the data gathering is instrumented there is a calibration bias of unknown amount so that the mean has no physical meaning with respect to the interpretation of the data. Therefore, the removal of the mean is a natural step to start with.

Now let us consider what the typical spectrum is. First, there is the true signal, usually confined to a reasonably narrow band of frequencies (the

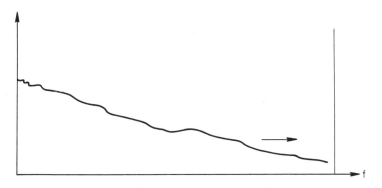

FIGURE 11.6-1 ORIGINAL NOISE SPECTRUM

reason will become apparent in a moment). Second, there is the noise, which by itself will generally have a spectrum that is gradually falling off as you go to infinity; the square of the area under the curve is often the total energy, and this is finite in all real systems. Therefore, it looks much like Figure 11.6-1. Next we remember that we have sampled the original signal and that this produces the aliasing (the folding back and forth), as shown in Figure 11.6-2. Of course, these frequencies are algebraically added, and there may be cancellation, but the result is that there is a strong tendency for the noise to have a "white" spectrum, to be fairly constant across the whole band of aliased frequencies. Thus the combination of signal plus noise is apt to look like Figure 11.6-3 where the signal corresponds to the wiggles.

Now you are in a position to filter out the noise where it is not in the same band as the signal; the narrower the signal band is with respect to the Nyquist interval, the relatively more of the noise you can remove. But what do you see when you first look? The mean is exactly the value of the spectrum at $k = 0$, and since in the complex form the spectrum is symmetric about

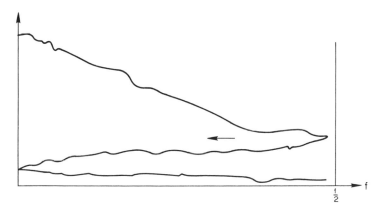

FIGURE 11.6-2 SPECTRUM FOLDED (ALASED)

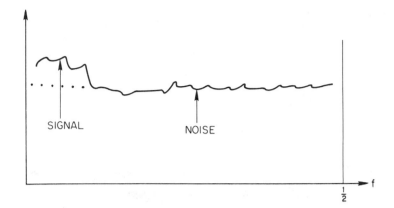

FIGURE 11.6-3 SIGNAL PLUS NOISE ALIASED

the origin, the removal of the mean makes the value at the origin exactly zero, thus putting a discontinuity into the spectrum! When you now use a von Hann or a Hamming window on the data, you raise this value up at the cost of lowering the adjacent values a bit. Thus the effect of the windowing causes you to see a spectrum that looks like Figure 11.6-4. The engineer is apt to say that there is a band of frequencies, not including the origin, in which the signal appears. But this is an artifact of the removal of the mean!

There is no simple answer to this problem. The removal of the mean greatly affects the appearance of the spectrum near the origin. All that I can advise is that, if you notice that the signal peaks at the third frequency point of the spectrum, then you should suspect that it is due to the removal of the mean.

Unfortunately, there is often a removal of more than the mean; the trend line, or even a trend parabola, may be removed before the spectrum is computed. Thus the Fourier coefficients you compute are the algebraic dif-

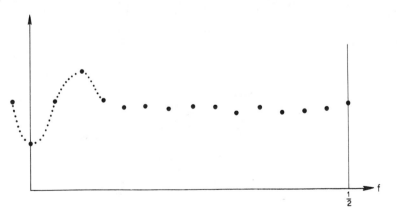

FIGURE 11.6-4 MEAN REMOVED, WINDOWED, AND SAMPLED

ferences between the true coefficients and the coefficients of the removed curve. The spectrum squares this difference, thus obscuring and making impossible any simple analysis. The removed part, generally having a dis continuity at the ends of the interval where it gets wrapped around the cylinder, has fairly large coefficients over the whole range of possible coefficients due to the falling off like $1/k$, plus the further aliasing due to the sampling. The situation is not pleasant to say the least.

All too often in practice this results in simply ignoring the whole set of phenomena that arise from the details of the processing and are not part of the underlying phenomena you are studying. The consistency of the results from run after run often reflects the consistency of the original data plus the consistency of how you are processing the data. Thus *the consistency of the interpretation of the results in no way implies its correctness.*

11.7 THE PHASE SPECTRUM

When you go from the Fourier coefficients to the power spectrum, you go from $2N$ coefficients (in either the real or complex form) to $N + 1$ terms. You lose effectively half the information. Occasionally, you see, along with the power spectrum, the corresponding *phase spectrum*, which, in some sense, contains the missing information. For real Fourier expansions the phase ϕ is given by

$$\cos \phi_k = \frac{a_k}{A_k}, \quad -\sin \phi_k = \frac{b_k}{A_k}, \quad A_k = \sqrt{a_k^2 + b_k^2}$$

(provided you remember that the quadrant that the phase falls in depends on the signs of both coefficients).

For the complex form of the Fourier series, we need to write the complex number

$$c_k = A_k e^{i\phi_k}$$

in the form of an exponential with pure imaginary value. We have, therefore (again watch the quadrant it falls in),

$$\phi_k = \arctan\left(-\frac{b_k}{a_k}\right)$$

In many problems the phase is important. We have seen that a square pulse requires that the Fourier terms that make it up must be properly aligned (in proper phase) so that in some regions they combine and in some regions they cancel. If such a pulse is sent down a transmission line whose delays differ for various frequencies, it is easy to see that at the receiving end there may not be a nice square pulse to detect. Similarly, if we allow phase relationships of the various frequencies to be different when we filter, then the nicely aligned terms may no longer combine properly. In some problems, increasingly so, only the phase is important.

Our simple nonrecursive filters, which were either symmetric for smoothing or antisymmetric for differentiators, kept all the phases in line with each other *when judged at the midpoint of the data* being processed at the moment. As we go on we will find that the phase looms more and more important in our understanding of what is happening.

11.8 SUMMARY

There are no simple answers to these problems. All the author can advise is that you be wary of any interpretation that is made from the spectrum and ask to what extent the methods of gathering and processing the data has made the effects that are seen. It is worth repeating the remark about the usual spectrum analysis. The $2N$ data points result in $2N$ coefficients, but in the spectrum almost half of these are lost. The corresponding phase spectrum has this lost information, but it is generally ignored. Why this ignoring of the phase spectrum which seems to contain almost half the information? It appears to the author that the reason is more tradition than common sense. It seems that the spectrum, indeed the whole frequency approach, first arose in the telephone business, where, for sound, the human ear is very insensitive to phase relations; the information in human speech is mainly in the amplitudes of the frequencies. They were justified in ignoring the phase spectrum. But many applications are now to other signals than speech, and often the phase has almost as much information as the amplitude. Indeed, there are applications where the phase contains all the information. Therefore, you should ask carefully about the justification of ignoring phase before you go down the usual path of concentrating on the power spectrum.

Thus we see that spectral analysis is a tricky topic. The preprocessing of the signal before you get it can, at times, greatly affect what you see. The statistician's customary removal of the mean and/or the trend line puts serious discontinuities into the spectrum along with all the aliasing and nonharmonic term effects. The topic is one that requires more than can be given in a first course that assumes only slight mathematical preparation, but we have at least warned you and provided you with simple arguments as to what to look for.

Ignoring all theory it is easy to make up selected sets of data to see what the proposed data processing of the signal will do, and often this is very convincing to the person who merely wants answers. What is required on your part is aggressive action in the direction of doubting that the results you get from the computations are what you think you are getting. At this point all we can recommend, besides taking more advanced courses, is a cautious, experimental approach to the topic of spectral analysis. The fact that the users get what they want is no test of the correctness of the results, nor of the validity of the data processing being done. It is all too easy to find the effects that you put in by how you do the processing and then to conclude that you are seeing reality rather than the artifacts of the methods used. Be skeptical!

12

Recursive Filters

12.1 WHY RECURSIVE FILTERS?

Equation (1.1-5) defined a recursive filter by the formula

$$y_n = \sum_{k=0}^{N} c_k u_{n-k} + \sum_{k=1}^{N} d_k y_{n-k} \qquad (12.1\text{-}1)$$

where in the second sum the $k = 0$ term is omitted. It is assumed, of course, that not all the d_k are zero; otherwise we would have a simple nonrecursive filter.

A recursive filter has more "memory" than a nonrecursive filter (all the $d_k = 0$); a nonrecursive filter can respond only to the values of u_m in the range u_{n-N} to u_{n+N}. An example of a recursive filter is a formula for integration that *must* "remember" all the past values back to and including the value at the lower limit of the integration. In Section 3.4 we examined the trapezoid rule for integration

$$y_{n+1} = y_n + \frac{h}{2}[u_{n+1} + u_n] \qquad (y_0 = \text{constant})$$

which is clearly a recursive filter. Similarly, Simpson's formula

$$y_{n+1} = y_{n-1} + \frac{h}{3}[u_{n+1} + 4u_n + u_{n-1}] \qquad (y_0 = \text{constant})$$

is a recursive filter.

230

Even a recursive filter with a short "span" of coefficients can remember all the past. In the simplest case,

$$y_{n+1} = ay_n + u_n \qquad (y_0 = 0)$$

where all the u_n are zero except $u_0 = 1$ and all the y_k for negative subscripts are zero, we get

$$y_1 = 1, y_2 = a, y_3 = a^2, y_4 = a^3, \ldots, y_n = a^{(n-1)}$$

Thus we have a geometric progression of values for y_n, and the value of u_0 persists indefinitely. This ability to remember values that occurred long ago represents a definite asset of the recursive filter in many situations.

However, a recursive filter may have only a finite memory. For example the filter

$$y_n = ay_{n-1} + \tfrac{1}{2}[u_n + (1 - a)u_{n-1} - au_{n-2}]$$

For a unit impulse of x_0 we have the following table:

n	u_n	y_n
-3	0	0
-2	0	0
-1	0	0
0	1	$\tfrac{1}{2}$
1	0	$\tfrac{1}{2}$
2	0	0
3	0	0

Many recursive filters are used in real-time problems, meaning that values of the data u_m and output y_m for subscripts $m > n$ are *not* available for use at the time that y_n is being computed. Such filters are often called (Section 1.1) *physically realizable*, as contrasted with the typical nonrecursive filter, which is *physically unrealizable*, meaning that when we calculate a value y_n, we are using data u_m for $m > n$ from the future. Although physically nonrealizable recursive digital filters are possible, they occur so seldom that we will concentrate on recursive filters of the form (12.1-1)

$$y_n = \sum_{k=0}^{M} c_k u_{n-k} + \sum_{k=1}^{N} d_k y_{n-k}$$

(but see Section 12.8). These should be called "one-sided" physically realizable filters, but we will use the name recursive filters for them. Often they are called *infinite impulse response* filters (IIR) from their ability to produce, from a single impulse, effects arbitrarily far into the future. Correspondingly, nonrecursive filters are abbreviated FIR, *finite impulse re-*

sponse. As we just showed, degenerate examples of recursive filters that do not continue an impulse indefinitely may be constructed but will not be considered here. Thus in some respects the name *infinite impulse response* filter is poorly chosen and is one of the reasons we prefer to use the name *recursive filter*.

Recalling that for a linear system the output frequency must be the same as the input frequency (Section 2.5), we make our usual substitutions

$$u_n = A_I e^{2\pi i f n}, \qquad y_n = A_O e^{2\pi i f n}$$

and obtain

$$\tilde{H}(f) = \frac{A_O}{A_I} = \frac{\displaystyle\sum_{k=0}^{M} c_k e^{-2\pi i f k}}{1 - \displaystyle\sum_{k=1}^{N} d_k e^{-2\pi i f k}} \qquad (12.1\text{-}2)$$

as our transfer function $\tilde{H}(f)$. Thus the transfer function of a recursive filter is a rational function in $e^{2\pi i f} = z$

If we write

$$e^{i\omega} = e^{2\pi i f} = z$$

then (12.1-2) becomes

$$\tilde{H}(f) = \frac{A_O}{A_I} = \frac{\displaystyle\sum_{k=0}^{M} c_k z^{-k}}{1 - \displaystyle\sum_{k=1}^{N} d_k z^{-k}}$$

We are now in a position to see why a recursive filter can have a second important property: the transition band from pass to stop can be narrow. This situation follows from the observation that, when the polynomial in the denominator goes near zero, the quotient can change rapidly, can rise or fall sharply, and allow a narrow transition band in a recursive filter.

The fact is a recursive digital filter, in mathematical terms, is merely a linear difference equation with constant coefficients; you are computing (12.1-1) and not (12.1-2).

The input terms, the u_{n-k}, form the *forcing term* of the equation

$$y_n - d_1 y_{n-1} - \cdots - d_n y_{n-m} = f(n) \equiv \sum c_k u_{n-k}$$

As a result of this, there are two distinct approaches to recursive digital filters. First, there is the mathematical approach through the theory of linear difference equations with constant coefficients (and this is reasonably familiar to the student who has had some linear differential equations with constant coefficients). The second approach goes directly to the transfer function and parallels the older approach to analog recursive filters. We shall use both approaches, particularly as the mathematical approach sheds a good deal of light on many of the details of the transfer function approach.

12.2 LINEAR DIFFERENTIAL EQUATION THEORY

The theory of linear difference equations almost exactly parallels the theory of linear differential equations with constant coefficients. We assume a slight familiarity with this theory, but will review it to remind the reader of some of the details. Treating the special case of second-order equations will not limit the understanding of the phenomena involved and makes everything much easier to understand.

We begin with the *complete equation*,

$$ay''(t) + by'(t) + cy(t) = f(t) \qquad (12.2\text{-}1)$$

and speak of the function $f(t)$ as the *forcing function*. Let $Y(t)$ be any particular solution of this equation. We will later say a few words about how to find it.

If we set the solution $y(t)$ equal to

$$y(t) = Y(t) + y_1(t)$$

then (12.2-1) becomes

$$a[Y''(t) + y_1''(t)] + b[Y'(t) + y_1'(t)] + c[Y(t) + y_1(t)] = f(t)$$

Since by definition

$$aY''(t) + bY'(t) + cY(t) = f(t)$$

upon subtraction we have

$$ay_1''(t) + by_1'(t) + cy_1(t) = 0$$

Thus knowing one solution of the complete equation (12.2-1) reduces us to solving the *homogeneous equation*

$$ay''(t) + by'(t) + cy(t) = 0 \qquad (12.2\text{-}2)$$

We next concentrate on the homogeneous equation

$$ay'' + by' + cy = 0$$

which is the same equation except that the right side is set equal to zero. Since it is a linear equation with constant coefficients we know a solution of the form (the eigenfunction)

$$y(t) = e^{mt}$$

Substitute this into the homogeneous equation. The result is

$$e^{mt}[am^2 + bm + c] = 0$$

Since the exponential cannot be zero for finite t, it follows that the *characteristic equation*

$$am^2 + bm + c = 0 \qquad (12.2\text{-}3)$$

must be satisfied by m. Let the two roots of this equation be

$$m_1 \quad \text{and} \quad m_2$$

For *each* m_i there is a solution of the homogeneous equation (12.2-3) of the form

$$C_i e^{m_i t}$$

(1) If the roots m_i are distinct, then there are two linearly independent eigenfunctions and the *general solution* of the homogeneous equation is

$$C_1 e^{m_1 t} + C_2 e^{m_2 t}$$

(2) If two of the roots are the same, then we try a solution of the form

$$t e^{mt}$$

and we will find that it is a solution. If in the general case of an nth order equation there were a triple root, then we would try both

$$t e^{mt} \quad \text{and} \quad t^2 e^{mt}$$

and so on. If there is a complex root $a + ib$ (where i is the $\sqrt{-1}$), then since the characteristic equation is real there is also the conjugate root $a -$

ib. Corresponding to these two roots we have

$$e^{(a+ib)t} \quad \text{and} \quad e^{(a-ib)t}$$

or else the equivalent pair of solutions

$$e^{at} \cos bt \quad \text{and} \quad e^{at} \sin bt$$

The theory for multiple complex roots follows the preceding pattern, where we put coefficients as powers of t into the expression we are trying.

In all cases we find that we have n linearly independent solutions of the homogeneous equation and that the general solution of the homogeneous equation (12.2-2) is a linear combination of these terms with arbitrary coefficients.

How do we find a particular solution $Y(t)$ of the complete equation? There is no simple method. If the forcing term on the right side, $f(t)$, is a polynomial of degree k, for example, then trying a general polynomial of degree k as the solution will, when you equate like powers of t on both sides, result in $k + 1$ linear equations whose solution defines the solution we are looking for. (But see later discussion.) Similarly, if the forcing term is an exponential, then we would try an exponential with the same exponent (provided it was not a characteristic root) and find the equation for its front coefficient C. In general, the form of the forcing term suggests in many cases the form of a solution to try. A warning is, however, necessary. If the solution you are trying is a solution of the homogeneous equation, then you need to raise the degree of the polynomial coefficient in front (a constant is a polynomial of degree 0). Suitably raising this degree, you will find a solution of the form if it is possible.

We need to observe that using this method the solution $Y(t)$ will not have any terms of the form of any of the terms of the homogeneous equation. It cannot happen! No part of the general solution of the homogeneous equation can be part of the particular solution of the complete equation using the *way* we have found it. Of course, any solution of the homogeneous equation can be added to the particular solution and it will automatically cancel out when the combined solution is put into the complete equation. But the way we have found that particular solution has ruled out any terms arising that are part of the solution of the homogeneous equation. We suppose that in any particular solution there are no pieces that belong to the homogeneous equation solution.

Example 1. Consider the equation

$$y'' + 4y' + 3y = 3t^2 \tag{12.2-4}$$

The homogeneous equation is

$$y'' + 4y' + 3y = 0$$

Trying the solution $y = e^{mt}$, we get the characteristic equation

$$m^2 + 4m + 3 = (m + 1)(m + 3) = 0$$

and the characteristic roots are $m_1 = -1$ and $m_2 = -3$. The general solution of the homogeneous equation (12.2-2) is therefore

$$y(t) = C_1 e^{-t} + C_2 e^{-3t}$$

We now need to find a particular solution of the complete equation (12.2-4). Since the right side is a polynomial of degree 2, we try

$$Y(t) = at^2 + bt + c$$

Put into the complete equation this gives the *identity*

$$3at^2 + (3b + 8a)t + (3c + 4b + 2a) = 3t^2$$

from which it follows (equating coefficients of like powers) that $a = 1, b = -\frac{8}{3}$, $c = \frac{26}{9}$. The general solution of the complete equation (12.2-4) is, therefore,

$$y(t) = C_1 e^{-t} + C_2 e^{-3t} + t^2 - \frac{8t}{3} + \frac{26}{9}$$

Example 2. As a second example, consider the equation

$$y'' + 2y' + y = e^{-t} \tag{12.2-5}$$

The homogeneous equation

$$y'' + 2y' + y = 0$$

leads to the characteristic equation

$$m^2 + 2m + 1 = (m + 1)^2 = 0$$

which has a double root -1. The solution of the homogeneous equation is therefore of the form

$$y(t) = C_1 e^{-t} + C_2 t e^{-t}$$

which can be tested if you wish.

To find a particular solution of the complete equation, you first think of the term

$$ae^{-t}$$

but that will not do since it is a solution of the homogeneous equation. The results are the same when you try this term multiplied by t. Thus you have to try

$$at^2 e^{-t}$$

You get

$$ae^{-t} \left[\begin{array}{l} \{t^2(-1)^2 + 4t(-1) + 2\} \\ + 2\{t^2(-1) + 2t\} + t^2 \end{array} \right] = e^{-t}$$

As expected the square and linear terms cancel out and the rest will yield $a = \frac{1}{2}$. Thus the general solution of the complete equation is

$$C_1 e^{-t} + C_2 t e^{-t} + (\tfrac{1}{2})t^2 e^{-t}$$

The next step in using the general solution of the complete equation is usually to fit the n initial conditions. These conditions determine the n coefficients C_i, (it can be shown that the algebraic equations are solvable in all cases).

If as t grows larger and larger the initial conditions are to have a smaller and smaller effect on the solution of the complete equation, then all the roots of the characteristic equation must have negative real parts. This is the classical definition of a stable system: the effect of the initial conditions gradually fades out with increasing t. However, if we are to handle digital filters that closely approximate integration, it is necessary to allow characteristic roots with real parts equal to 0. Indeed, if we are later to handle the numerical integration of a second-order differential equation with no first derivative, then we must allow a double root of 0. Thus *we must allow* the solution of the homogeneous equation to be a constant or even grow like a polynomial in t. However, unless there is an exponential growth in the forcing function $f(t)$ we do not need to allow exponential growth in the solution of the homogeneous equation. Thus we are forced to a new definition of stability. *The system is said to be stable if the real parts of the roots of the characteristic equation are not positive.*

We might have kept the old definition of stable and used words like "quasi-stable" or "conditionally stable," but after careful consideration it seems best to face up to the changed circumstances of having digital computers and appropriately alter the definition. It is only in this way that theories can evolve to fit new situations.

12.3 LINEAR DIFFERENCE EQUATIONS

We now turn to the corresponding theory of linear difference equations

$$ay_n + by_{n-1} + cy_{n-2} = f(n) \qquad (12.3\text{-}1)$$

with constant coefficients. In the theory it is convenient to write

$$e^{mt} \text{ in the form } \rho^n \qquad (12.3\text{-}2)$$

(recall that n plays the role of t). Otherwise, things are exactly the same. Here ρ plays the role of e^m. Upon direct substitution of the assumed solution into the equation we are led to the characteristic equation

$$a\rho^2 + b\rho + c = 0 \qquad (12.3\text{-}3)$$

For multiple roots we try $n\rho^n$, $n^2\rho^n$, and so on. Everything goes the same otherwise. Stability becomes not the condition on the real part of m that it be nonpositive, but rather that ρ be *not greater than 1 in size*. The condition $m = 0$ corresponds to $\rho = 1$.

Example 3. Consider the difference equation

$$y_n - 4y_{n-1} + 3y_{n-2} = 4n \qquad (12.3\text{-}4)$$

The characteristic equation of the homogeneous equation is

$$\rho^2 - 4\rho + 3 = (\rho - 3)(\rho - 1) = 0 \qquad (12.3\text{-}5)$$

and hence the general solution of (12.3-4) is

$$C_1(3)^n + C_2(1)^n = C_1(3)^n + C_2$$

which is not stable. The particular solution we search for is of the form of a polynomial in n (since 1 is a characteristic root we must raise the degree by 1, hence in place of $an + b$ we must try)

$$an^2 + bn$$

This leads to the identity in n of the form

$$an^2 + bn - 4[a(n-1)^2 + b(n-1)] + 3[a(n-2)^2 + b(n-2)] = 4n$$

Equating like powers

$$n^2[a - 4a + 3a] = 0$$

$$n\lfloor b + 8a - 4b - 12a + 3b] = 4n$$

$$[-4a + 4b + 12a - 6b] = 0$$

Hence

$$-4a = 4$$

$$8a - 2b = 0$$

from which we can obtain the coefficients $a = -1$, $b = -4$.

Example 4. Next consider the equation

$$y_n - 2y_{n-1} + y_{n-2} = f(n)$$

The characteristic equation is

$$\rho^2 - 2\rho + 1 = 0$$

and has multiple roots so that the solution is

$$y(n) = C_1(1)^n + C_2 n(1)^n = C_1 + C_2 n$$

and this, by our revised definition, is stable.

In this case where there is a double root at 1, if $f(n)$ has the form of a constant or is linear, we will have to try forms with a highest power of n, two (multiplicity of root at unity) higher than the highest power of n in $f(n)$, to get the particular solution, but that is not of interest here.

Thus we see that the theory of linear difference equations almost exactly parallels the theory of linear differential equations, and there is little new to learn. The chief difference is the use of ρ rather than the exponential e^m. Stability becomes the condition that the characteristic roots in ρ must not be bigger than 1 in size,

$$|\rho| \leq 1$$

The growth of the solution of the homogeneous equation (through which the initial conditions enter) may be a polynomial if there are multiple roots, but no greater a growth is allowed.

Remember again it is this difference equation that you use to compute; indeed *the equation is the digital filter.*

12.4 REDUCTION TO SIMPLER FORM

We now develop the conventional transfer function approach to recursive digital filters.

Recall that in order to get smooth filters (Chapter 7), we made a transformation that converted the problem into one that was equivalent to fitting a polynomial to the transfer function characteristic that we wanted (Section 7.2). In the recursive filter case, we do not have a polynomial but rather a rational function in both sines and cosines. From calculus, we remember that a rational function in both sines and cosines can be reduced to a rational function in w by the *tangent half-angle substitution*

$$\tan \frac{\omega}{2} = \tan \frac{2\pi f}{2} = w$$

Before using this transformation blindly, let us pause and look a little closer to how and why it works. Actually, we are in the position of having a rational function in $e^{2\pi i f}$.

When we have a polynomial in w, the most general "one-to-one" substitution that we can make for w and still preserve the polynomial form of the same degree is a linear substitution:

$$w' = aw + b, \qquad w = \frac{1}{a}(w' - b)$$

However, with a rational function we have an increased range of substitutions that will not alter the basic form and complexity; we can now use the linear fractional transformation, which is still one-to-one,

$$w' = \frac{aw + b}{cw + d}, \qquad w = \frac{dw' - b}{a - cw'} \tag{12.4-1}$$

Notice that one of the coefficients can be taken to be 1.

This transformation is also called the *bilinear transformation*. It is also sometimes called *frequency warping*, since it is a nonlinear transformation of the frequency.

We first ask the question: Which transformations in this three-parameter family

$$z = e^{2\pi i f} = \frac{aw - b}{cw + d} \tag{12.4-2}$$

will have the property that for real f we get real w? We can determine the details of the transformation as follows.

To fix the *phase* of f (which otherwise would be arbitrary and only add to the confusion), we can require of (12.4-2) that

$$f = 0 \qquad \text{corresponds to} \qquad w = 0$$

This means

$$e^0 = 1 = \frac{b}{d} \qquad \text{or} \qquad b = d$$

Since one coefficient is arbitrary, we can set

$$b = d = 1$$

We now have (12.4-2) in the form

$$z = e^{2\pi i f} = \frac{1 + aw}{1 + cw} \tag{12.4-3}$$

To determine the *direction* of the transformation, we can require

$$f \rightarrow \tfrac{1}{2} \text{ corresponds to } w \rightarrow + \infty$$

This means that at $f = \tfrac{1}{2}$ we have from (12.4-3)

$$e^{\pi i} = -1 = \frac{a}{c} \qquad \text{or} \qquad c = -a$$

We now have

$$z = e^{2\pi i f} = \frac{1 + aw}{1 - aw} \tag{12.4-4}$$

Finally, to fix the *scale* of the transformation, we can require for (12.4-4)

$$f = \tfrac{1}{4} \qquad \text{corresponds to} \qquad w = 1$$

This means

$$e^{\pi i/2} = i = \frac{1 + a}{1 - a}$$

Multiply both sides by $(1 - a)$ to obtain

$$i - ia = 1 + a$$

which imples

$$a = i$$

We have finally determined the transformation completely (using the three conditions):

$$z = e^{2\pi i f} = \frac{1 + iw}{1 - iw} \qquad (12.4\text{-}5)$$

Notice that for real f and real w each side has absolute value 1.
    Writing this formula in the *real* form

$$\cos 2\pi f + i \sin 2\pi f = \frac{1 + 2iw - w^2}{1 + w^2}$$

and equating real and imaginary parts gives the usual *tangent half-angle* formulas of the calculus

$$\cos 2\pi f = \frac{1 - w^2}{1 + w^2}, \qquad \sin 2\pi f = \frac{2w}{1 + w^2} \qquad (12.4\text{-}6)$$

As stated in the first paragraph of this section,

$$\tan \pi f = \tan\left[\frac{2\pi f}{2}\right] = \frac{1 - \cos 2\pi f}{\sin 2\pi f} = \frac{1 + w^2 - (1 - w^2)}{2w} = w \quad (12.4\text{-}7)$$

Using this transformation, our transfer function becomes a rational function in w with complex coefficients.
    Before continuing, let us examine the transformation (Figure 12.4-1). As the original variable f went from 0 to $\frac{1}{2}$, the new variable w goes from 0 to infinity. Because of symmetry, as f goes from 0 to $-\frac{1}{2}$, w goes from 0 to minus infinity.
    This transformation is often called "frequency warping" but it is simply the standard tangent half-angle substitution of conventional calculus.

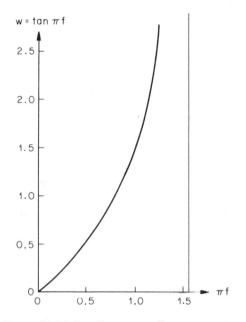

FIGURE 12.4-1 THE FREQUENCY TRANSFORMATION

For our transfer function of the recursive filter, we now have a rational function in w, with w real, but we still have complex coefficients. It is conventional to ignore the effects due to the imaginary terms, the phase relation (for each frequency) between the input and the output of the filter; in mathematical words, we ignore the actual complex value of the ratio of A_0/A_I and concentrate on its absolute value. This step is easiest done by taking the product of $H(f)$ times its conjugate (conjugate means replace i by $-i$). The product is, of course, real. Since i occurs only with f in the expressions, the procedure is the same as using

$$\tilde{H}(f)\tilde{H}(-f) \tag{12.4-8}$$

To summarize, we have replaced the original problem, that of finding a recursive filter to fit a given transfer function in the Nyquist interval, with the problem of matching the square of the modulus of the transfer function along the whole real axis w. The latter *omits all information about the phase relationships* between the input and output signal frequencies (but see Section 12.8). Using the tangent half-angle transformation and (12.4-8), we now have a problem of fitting a rational function in w, over the range $-\infty$ to $+\infty$, to the square of the modulus of the transfer function (transformed, of course, to the w variables). The step (12.4-8) will ultimately have to be undone.

12.5 STABILITY AND THE Z TRANSFORMATION

Before going deeper into the design problem, let us look at the problem of undoing the replacement of the original problem of matching $\tilde{H}(f)$ by the problem of matching the product

$$\tilde{H}(f)\tilde{H}(-f) \qquad (12.5\text{-}1)$$

This replacement has doubled the degree of the rational function and, in particular, has doubled the number of zeros in the denominator. As is always the case in computing when using old values, we need to consider the *stability* of the digital filter. By "stability" we usually mean that a bounded sequence of input values will produce a bounded output, although not necessarily the same bound. Actually, we found it necessary to allow a polynomial growth to the output. In short, we must return to our original difference equation and see what restriction must be placed on it in order to achieve this stability. When we know this, then we will have some idea of how to go from the new problem back to our original problem.

We write our difference equation in the form

$$y_n = \sum_{k=1}^{N} d_k y_{n-k} + \sum_{k=0}^{M} c_k u_{n-k}$$

The u_{n-k} are the given data, and so the last sum can be regarded as a known function of n. We are therefore reduced to studying the simple linear difference equation in y_n with a forcing function (the second sum).

We used ρ a moment ago because it is conventional notation in the field of difference equations, but in recursive filter theory it is customary to use z, where

$$\rho = z = e^{2\pi i f}$$

We now make a major shift in our thinking for z is no longer necessarily on the unit circle and the variable f is not necessarily real.

We are looking for solutions of the difference equation of the form

$$z^m$$

The characteristic polynomial of the difference equation (12.1-1) of the filter is then the polynomial

$$z^N - \sum_{k=1}^{N} d_k z^{N-k} = 0 \qquad (12.5\text{-}2)$$

(which is the denominator of the transfer function) and the condition that no solution of this polynomial be greater than 1 in size means that no solution of the homogeneous equation can grow exponentially. The classical analog filter theory at this point requires that all the zeros be less than 1; but since we are interested in numerical integration, among other things, we must permit a zero on the boundary of the circle. For second-order linear differential equations without a first derivative present, we must allow a double zero (which leads to a solution of the homogeneous difference equation that grows like n, the number of steps integrated).

The solution of the homogeneous equation is often called the *transient* solution. Stated simply, the conventional theory requires that the transient die out sufficiently far away, but we are permitting it to persist (so we can remember the initial conditions for integration, for example), and, indeed, in some cases allow polynomial growth, but we still avoid exponential growth in the homogeneous equation.

The transformation from the z variable to the w variable is

$$z = \frac{1 + iw}{1 - iw} \quad \text{or} \quad w = i\left[\frac{1 - z}{1 + z}\right] \tag{12.5-3}$$

In the w notation, the stability condition $|z| \leq 1$ becomes

$$|z| = \left|\frac{1 + iw}{1 - iw}\right| \leq 1$$

To apply this condition, we first recall that z goes around the unit circle in the complex plane from $-\pi$ to $+\pi$ (Figure 12.5-1). On the circle we have

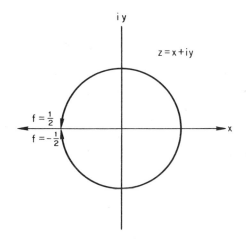

FIGURE 12.5-1 THE z TRANSFORMATION

$z = e^{2\pi i f}$ for real f. Correspondingly, w goes along the real axis from $-\infty$ to $+\infty$, and we see that the inside of the unit circle in the z plane is mapped into a half-plane in w. To decide which half, we simply find where $z = 0$ goes. From (12.5-3) this point is $w = i$. Thus the inside of the circle $|z| = 1$ goes into the upper half of the w plane. For our filters we will therefore require that the zeros of the denominator of the transfer function all lie in, or on the edge of, the upper half plane in w. (Some textbooks use a mapping of z to w such that it is the right or left half-plane that can be used, which is only a question of an extra multiplier of i in the transformation.)

The problem of selecting which zeros of the denominator to use for a stable filter arises from the fact that we were given a transfer function $\tilde{H}(f)$, and to ensure that it was real, we started our design by approximating (12.5-1)

$$\tilde{H}(f)\tilde{H}(-f)$$

Since f only occurs in conjunction with i, this expression is equivalent, as we saw, to the product of the transfer function times its complex conjugate. When we finally derive the approximation to this real function, we face the messy problem of extracting a suitable $\tilde{H}(f)$ that will be stable (zeros all in the upper half-plane or on the edge).

12.6 BUTTERWORTH FILTERS

Perhaps the most famous recursive filter design is the Butterworth filter. Such filters correspond to our earlier smooth nonrecursive filters (Chapter 7). For the Butterworth filter we simply pick the rational function

$$\tilde{H}(f)\tilde{H}(-f) = \frac{1}{1 + [w/w_c]^{2N}} \tag{12.6-1}$$

This function has the obvious features of high tangency at both the origin and infinity (in the w variable) and smoothness elsewhere. (Compare with Chapter 7.)

The design features are shown in Figure 12.6-1. Based on this figure, we can obtain the design parameters, N and w_c, as follows. First, we write the two conditions that arise at the edges of the transition band,

$$\frac{1}{1 + \epsilon^2} = \frac{1}{1 + [w_p/w_c]^{2N}}$$

$$\frac{1}{A^2} = \frac{1}{1 + [w_s/w_c]^{2N}}$$

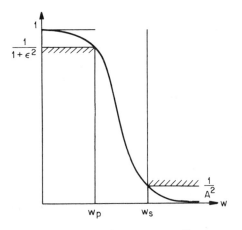

FIGURE 12.6-1 BUTTERWORTH FILTER

Next, take the reciprocal of both equations, do some simplification, and get

$$\left[\frac{w_p}{w_c}\right]^{2N} = \epsilon^2$$

$$\left[\frac{w_s}{w_c}\right]^{2N} = A^2 - 1 \qquad (12.6\text{-}2)$$

To eliminate w_c, divide one equation by the other

$$\left[\frac{w_p}{w_s}\right]^{2N} = \frac{\epsilon^2}{A^2 - 1}$$

Solving for N, we have

$$N = \frac{\log{(\epsilon/\sqrt{A^2 - 1})}}{\log{(w_p/w_s)}} \qquad (12.6\text{-}3)$$

and hence from the top equation of (12.6-2)

$$w_c = \frac{w_p}{(\epsilon)^{1/N}} \qquad (12.6\text{-}4)$$

Thus we have determined the product $\tilde{H}(f)\tilde{H}(-f)$.

To find a suitable $\bar{H}(f)$, we proceed as follows. The zeros of the denominator of (12.6-1) are

$$1 + \left[\frac{w}{w_c}\right]^{2N} = 0 \tag{12.6-5}$$

$$\left[\frac{w}{w_c}\right]^{2N} = -1 = e^{\pi i + 2\pi i k}$$

$$\frac{w}{w_c} = e^{\pi i(2k+1)/2N} \qquad (k = 0, 1, \ldots, 2N - 1)$$

Of these $2N$ zeros, the zeros that we want are in the upper half-plane and are clearly given by

$$k = 0, \ldots, N - 1 \tag{12.6-6}$$

If we pair these zeros off one from each end, that is, pair k with $N - k - 1$, we obtain for each pair

$$\left[\frac{w}{w_c} - e^{\pi i(2k+1)/2N}\right]\left[\frac{w}{w_c} - e^{\pi i(2N-2k-1)/2N}\right] \tag{12.6-7}$$

Since

$$e^{\pi i(2N/2N)} = e^{\pi i} = -1$$

(12.6-6) becomes

$$\left[\frac{w}{w_c} - e^{\pi i(2k+1)/2N}\right]\left[\frac{w}{w_c} + e^{-\pi i(2k+1)/2N}\right]$$

$$= \left[\frac{w}{w_c}\right]^2 - 2i \sin\left\{\frac{\pi(2k + 1)}{2N}\right\}\left[\frac{w}{w_c}\right] - 1$$

To get back to the z variable, we substitute (12.5-3)

$$w = i\left[\frac{1 - z}{1 + z}\right]$$

and we have

$$\left[\frac{-(1/w_c^2)(1 - z)^2 + 2\sin\{\pi(2k + 1)/2N\}[1 - z^2][1/w_c] - (1 + z)^2}{(1 + z)^2}\right]$$

$$\tag{12.6-8}$$

as the real quadratic factor corresponding to the two complex, conjugate linear factors. There are, of course, at most $N/2$ of these factors, one for each k, $0 \leq k \leq (N/2)$.

If N is odd, then there is one factor left,

$$\left[\frac{w}{w_c} - e^{\pi i/2} \right] = \left[\frac{w}{w_c} - i \right]$$

and the substitution back to z gives

$$i \left[\frac{(1/w_c)(1 - z) - (1 + z)}{1 + z} \right]$$

These factors were in the denominator; thus inverting them gives the product

$$\tilde{H}(f) = \prod \frac{(1 + z)^2}{\{\text{quadratic factors}\}}$$

with a possible linear factor included. The construction of a polynomial from its zeros leaves the multiplier of the whole polynomial arbitrary. We use the fact that at $f = 0$ we want unity gain in the filter, which means that we must drop the multiplier of -1 in each quadratic factor and i in the linear factor *when* it occurs in the reconstruction of the polynomial from its zeros.

It might be supposed that at this point we would multiply out the numerator and denominator to find the coefficients of the individual powers of z and thus obtain the coefficients of the filter. However, doing so is unnecessary *and* can lead to serious roundoff errors. Instead we will simply pair off a factor of degree 2 in the numerator with each quadratic factor in the denominator and a factor of degree 1 if there is a final linear factor in the denominator. In this form we have the desired filter transfer function characteristic $\tilde{H}(f)$ as a cascade (product) of second-order transfer functions (possibly with one first-order) (remember we dropped a minus sign).

$$\tilde{H}_c(f) = \frac{w_c^2[z^2 + 2z + 1]}{z^2[w_c^2 + 2w_c \sin\{\pi(2k + 1)/2N\} + 1]} \qquad (12.6\text{-}9)$$
$$- z[2 - 2w_c^2] + [w_c^2 - 2w_c \sin\{\pi(2k + 1)/2N\} + 1]$$

where

$$z = e^{2\pi i f}$$

is the standard z-transform.

To get back to the original digital filter, recall that the second-order recursive digital filter (12.1-1)

$$y_n = c_0 u_n + c_1 u_{n-1} + c_2 u_{n-2} + b_1 y_{n-1} + b_2 y_{n-2}$$

has (Section 12.1) the transfer function

$$\tilde{H}_c(f) = \frac{c_0 + c_1 z^{-1} + c_2 z^{-2}}{1 - b_1 z^{-1} - b_2 z^{-2}} = \frac{c_0 z^2 + c_1 z + c_2}{z^2 - b_1 z - b_2} \quad (12.6\text{-}10)$$

Thus we can identify the coefficients of the transfer function with the coefficients of the digital filter that we want: the process requires noting that the coefficient of z^2 in the denominator of the second form is 1, and so in the first form we must divide each coefficient by the coefficient of z^2 before equating the corresponding coefficients of the two forms.

Our original data u_n is now processed by one second-order filter; the output is processed by the next filter, and so on, perhaps including one first-order filter. In this way, we have the results wanted, the original u_n processed by the Nth-order Butterworth filter.

This pairing off of the conjugate roots and the construction of the corresponding second-order filter will be used repeatedly—hence it is important to understand what is going on.

12.7 A SIMPLE CASE OF BUTTERWORTH FILTER DESIGN

To illustrate the design of a Butterworth filter, we select the simplest example (smallest N) of a Butterworth filter that illustrates the method adequately and carry out the numerical details carefully, following the steps in the previous section. Let $N = 3$. First, we find the zeros of

$$1 + \left(\frac{w}{w_c}\right)^6 = 0$$

which are

$$\frac{w}{w_c} = e^{\pi i(1 + 2k)/6}$$

Only for $k = 0, 1, 2$ (see 12.6-6) does the angle satisfy

$$0 \le \frac{\pi(1 + 2k)}{6} \le \pi$$

which is the condition for the value to be in the upper half-plane. We pair the cases $k = 0$ and 2 to form the quadratic factor (12.6-8)

$$
\left[\frac{w}{w_c} - e^{\pi i/6}\right]\left[\frac{w}{w_c} - e^{5\pi i/6}\right] = \left[\frac{w}{w_c} - e^{\pi i/6}\right]\left[\frac{w}{w_c} + e^{-\pi i/6}\right]
$$

$$
= \left(\frac{w}{w_c}\right)^2 - \frac{w}{w_c}\left[e^{\pi i/6} - e^{-\pi i/6}\right] - 1
$$

$$
= \left(\frac{w}{w_c}\right)^2 - 2i\,\frac{w}{w_c}\,\sin\frac{\pi}{6} - 1
$$

But $\sin \pi/6 = \frac{1}{2}$, and so we have

$$
\left(\frac{w}{w_c}\right)^2 - i\,\frac{w}{w_c} - 1
$$

The remaining, unpaired factor comes from $k = 1$:

$$
\frac{w}{w_c} - e^{\pi i/2} = \frac{w}{w_c} - i
$$

We convert these factors from w to z, using

$$
w = \frac{i(1 - z)}{1 + z}
$$

and get (dropping the multiplicative factor of -1 in the quadratic term and i in the first-order filter (12.6-9) so that at $w = 0$ we have unity gain)

$$
\frac{1}{\left[\left(\dfrac{1-z}{1+z}\right)^2 \dfrac{1}{w_c^2} + e^{\pi i}\left(\dfrac{1-z}{1+z}\right)\dfrac{1}{w_c} + 1\right]\left[\left(\dfrac{1-z}{1+z}\right)\dfrac{1}{w_c} - 1\right]}
$$

$$
= \left[\frac{(1 + z)^2 w_c^2}{z^2(1 + w_c^2 + w_c) + 2z(w_c^2 - 1) + (1 + w_c^2 - w_c)}\right]
$$

$$
\times \left[\frac{(1 + z)w_c}{-1 + w_c + z(1 + w_c)}\right] = \tilde{H}(f)
$$

which is the product of a second-order section and a first-order section filter. These two simple filters are then used in cascade (one filter feeding into the next one) to process the original signal x_n.

12.8 REMOVING THE PHASE: TWO-WAY FILTERS

Butterworth filters (and most other recursive filters) have, since they are not symmetric, a phase relation between the input and output signals that is not the same for all frequencies. Often this phase is annoying, but *if* we can process the data in both a forward and a backword direction, then there is a simple solution to the problem of eliminating the phase. We merely process the data by the linear filter and run this output *taken in the reverse direction* through the same filter. If there is a phase shift at a given frequency in the first pass through the filter, there is the same phase shift of the opposite sign at this same frequency in the second pass. Because we are processing the output of the first pass in the reverse direction, the two phase shifts must exactly cancel. We, of course, obtain the *square* of the absolute value of the transfer function as the effective transfer function. Thus, if we wish to use this trick, then in our original design we allow for this squaring of the transfer function that arises from the two passes.

The trick is so simple that it is apt to be overlooked. The reader is reminded especially to keep in mind those situations in which the data may be processed in either of the two directions. Typically, this procedure can be done if the data is fully recorded before the processing starts. It clearly *cannot* be done in real-time data processing. Remember that the double pass will square the transfer function.

13

Chebyshev Approximation and Chebyshev Filters

13.1 INTRODUCTION

The problem of measuring how close a designed filter is to the original ideal transfer function that we started with occurs constantly. We have used the measure of least squares, modifying it whenever we applied a window; in this chapter we introduce another criterion of closeness, the *Chebyshev criterion*, which uses the maximum difference between two curves as the measure of the distance between them.

The Chebyshev criterion is very popular these days, not only in filter theory where we wish to know the worst that we have done, but also in many situations in computing where, for instance, we wish to know that a certain library routine gives answers "at least as good as. . . ." In finding a Chebyshev fit, we adjust the parameters to make the maximum error as small as possible: we *minimize the maximum error*, and so the process is often called the *minimax strategy* of fitting.

The price to be paid for this fit, as compared to least squares, is that the sum of the squares of the errors is larger for a Chebyshev fit than for the corresponding least-squares fit; but also, conversely, the least-squares fit has a larger maximum error than does the Chebyshev fit. In filter design the Chebyshev fit is frequently preferred. Almost always the Chebyshev criterion distorts the least-squares fit less than least squares distorts the Chebyshev.

In the Chebyshev approximation we usually have an error curve that turns out to have equal-sized ripples. The reason is intuitively evident. Imagine that you have some kind of fit to the data and that you are looking at the error curve. If you try to push down the worst peak of the error curve by a change of the filter coefficients, then some other place will naturally increase in size a little. If at any stage you try to push down all the worst extremes at the same time, you will (probably) be able to do so until there are more maximum-sized extremes than you have adjustable parameters, whereupon you can go no farther. Thus the Chebyshev approximation is often called the *equal ripple approximation*.

Our approach to the design of Chebyshev filters will be via the Chebyshev polynomials, which individually have the equal ripple property and are therefore closely related to the equal ripple fitting process. Finally, we will look at the Chebyshev design of a simple integrator that is *not* based on the transfer function approach.

13.2 CHEBYSHEV POLYNOMIALS

The Chebyshev polynomials are defined by

$$T_n(x) = \cos(n \arccos x) \qquad (n = 1, 2, \ldots)$$
$$T_0(x) = 1 \qquad\qquad (13.2\text{-}1)$$

As the definition stands $T_n(x)$ does not look like a polynomial of degree n. However, in Section 7.2 we showed that $\cos nx$ is a polynomial in $\cos x$ of exactly degree n. We now see, since $\cos(\arccos x) = x$, that our definition of the Chebyshev polynomial $T_n(x)$ does indeed define a polynomial of degree n.

The Chebyshev polynomials are closely related to the cosine functions, as can be seen from the following change of variable

$$x = \cos \tau \qquad\qquad (13.2\text{-}2)$$

or, equivalently,

$$\tau = \arccos x \qquad\qquad (13.2\text{-}3)$$

See Figure 13.2-1 for what the transformation does in detail, how it stretches the independent variable when going from one notation to the other (equally spaced τ do not give equally spaced x). It is convenient to define the polynomials for negative subscripts by the obvious formula

$$T_{-k}(x) = T_k(x)$$

From the fact that the cosines are orthogonal functions (over both the

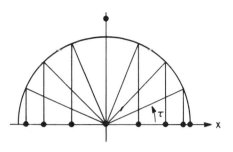

FIGURE 13.2-1

continuous interval and discrete sets of points), we see that the continuous Chebyshev polynomials are also orthogonal with some weight function over the interval $-1 \le x \le 1$. To find the details, we start with the orthogonality conditions of Section 4.2 (note that as a result of Section 10.2 there are also discrete versions corresponding to the continuous version that we are using):

$$\int_0^\pi \cos m\tau \cos n\tau \, d\tau = \begin{cases} 0, & m \ne n \\ \dfrac{\pi}{2}, & m = n \ne 0 \\ \pi, & m = n = 0 \end{cases}$$

Changing the variable from τ to x, using $\tau = \text{arc cos } x$, we get

$$\int_{-1}^1 \frac{T_m(x)T_n(x)}{\sqrt{1 - x^2}} \, dx = \begin{cases} 0, & m \ne n \\ \dfrac{\pi}{2}, & m = n \ne 0 \\ \pi, & m = n = 0 \end{cases} \qquad (13.2\text{-}4)$$

From this result we see that the Chebyshev polynomials are orthogonal over the interval $-1 \le x \le 1$ with weight function

$$\frac{1}{\sqrt{1 - x^2}}$$

For the discrete set of points, the unequal spacing of the sample points plays the role of the weight function in the summation over the sample points to produce the orthogonality.

From the trigonometric identity

$$\cos(n + 1)\tau + \cos(n - 1)\tau = 2 \cos \tau \cos n\tau$$

we obtain, by changing variables from τ to x (13.2-2) and using (13.2-1), the three-term recurrence relation

$$T_{n+1}(x) - 2xT_n(x) + T_{n-1}(x) = 0 \qquad (13.2\text{-}5)$$

It is now easy to generate the first few Chebyshev polynomials [clearly, $T_0(x) = 1$ and $T_1(x) = x$], using (13.2-5)

$$T_2(x) = 2x^2 - 1$$
$$T_3(x) = 4x^3 - 3x$$
$$T_4(x) = 8x^4 - 8x^2 + 1$$

$$\cdots$$

An examination of these polynomials and the three-term relation (13.2-5) shows that the nth polynomial has a leading coefficient of

$$2^{n-1}$$

and that the polynomials are alternately odd and even functions of x. Examination also shows that since $T_0(1) = T_1(1) = 1$, it follows that $T_n(1) = 1$ for all n.

Exercises

13.2-1 Prove $T_n(-1) = (-1)^n$.

13.2-2 Compute $T_5(x)$ and $T_6(x)$.

13.2-3 Show that $T_{2n}(x) = T_n(2x^2 - 1) = 2T_n^2(x) - 1$.

13.2-4 Show that $T_n[T_m(x)] = T_{mn}(x)$.

13.3 THE CHEBYSHEV CRITERION

The theorem that we prove in this section, and the one that connects Chebyshev polynomials with the minimax criterion, is that of all polynomials of degree n with leading coefficient 1 the Chebyshev polynomial divided by 2^{n-1} has the least maximum in the interval $-1 \le x \le 1$ [$T_n(x)/2^{n-1}$ is the "smallest" polynomial of degree n beginning $x^n + \cdots$].

The proof is fairly easy. To start, we note that the nth Chebyshev polynomial, being a cosine in disguise, has the equal ripple property and has exactly $n + 1$ extreme values in the interval. If there were another polynomial, say $P_n(x)$, with smaller extreme values, then the difference

$$\frac{T_n(x)}{2^{n-1}} - P_n(x) = \text{a polynomial in } x \text{ of degree at most } (n-1)$$

We now consider the value of this expression at the $(n+1)$ extreme values of the nth Chebyshev polynomial. Clearly, by the definition of $P_n(x)$, the values of this difference will be alternately positive and negative at successive extremal points. (See Figure 13.3-1.) Thus there will be at least n changes of sign (and hence n zeros) for the difference of the functions which is a polynomial of degree at most $(n-1)$, a contradiction! We conclude, therefore, that the Chebyshev polynomial divided by 2^{n-1} has the smallest value in the interval of any polynomial of that degree with leading coefficient 1.

This theorem shows the central role of the Chebyshev polynomials when we are interested in the minimax (Chebyshev) criterion for selecting a polynomial fit. If the error curve can be written as a simple Chebyshev polynomial, then we will have the best Chebyshev fit. If we can represent a function as a series in Chebyshev polynomials and truncate the series, the error will look like the first neglected term, *provided* the series is reasonably rapidly convergent.

The fact that the Chebyshev polynomials are also orthogonal polynomials means, as we have seen, that they give the best least-squares fit (relative to their weight function) in the interval. That they also give the minimax fit seems to be a contradiction. It is the weight function

$$\frac{1}{\sqrt{1 - x^2}} \qquad (-1 \le x \le 1)$$

which puts great weight at the ends of the interval where the least-squares fit tends to be worst, that explains how an expansion in Chebyshev poly-

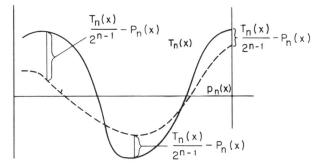

FIGURE 13.3-1

nomials can give both the best-weighted least-squares fit and *almost* the best minimax fit (the error is actually not the first term neglected but the sum of all the neglected terms). Note that an expansion in Chebyshev polynomials is a Fourier cosine expansion in disguise.

13.4 CHEBYSHEV FILTERS

In the chapters on nonrecursive filters we saw the advantage, in terms of a narrower transition band, of allowing ripples in either the passband or stopband of a filter, or in both. So we now turn to low-pass *recursive* filters in which we allow ripples in either the passband or stopband.

The criterion used in our design is the Chebyshev criterion of the minimum-maximum error, the equal ripple approach. This choice is quite natural in view of the use that we usually make of a digital filter; we often want to control the *maximum* error.

The design method is, as in the case of the Butterworth filters, "Behold! this formula does the trick." Thus we need only observe that the proposed formula has the properties that it is supposed to have; no elaborate theory is necessary to deduce the design or to use later if something different but quite similar to the standard design objective is wanted.

13.5 CHEBYSHEV FILTERS, TYPE 1

The first design, type 1, allows ripples in the passband (Figure 13.5-1). To meet this design, we start with (in the polynomial variable w of Section 12.4)

$$\tilde{H}(f)\tilde{H}(-f) = \frac{1}{1 + \epsilon^2[T_n(w/w_p)]^2} \qquad (13.5\text{-}1)$$

where $T_n(x)$ is the Chebyshev polynomial of order n. We see immediately that as long as $|w| \leq |w_p|$, we have

$$\frac{1}{1 + \epsilon^2} \leq |\tilde{H}(f)\tilde{H}(-f)| \leq 1$$

and at $w = w_p$ we have the equal ripple Chebyshev polynomial beginning to break out of the limits $-1 \leq T(w/w_p) \leq 1$.

The fact that the Chebyshev polynomial in the interval $-1 \leq x \leq 1$ is the smallest of any polynomial of the same degree and same leading coef-

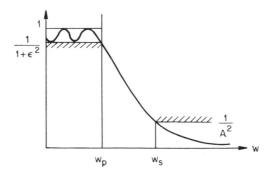

FIGURE 13.5-1 CHEBYSHEV TYPE 1 FILTER

ficient means, conversely, that given the bound, it rises most rapidly outside (we are sure of this fact if we go far enough away from 1, but it is fairly obvious that it applies near 1).

Recall that

$$\sin x = \frac{e^{ix} - e^{-ix}}{2i}, \qquad \sinh x - \frac{e^x - e^{-x}}{2}$$

$$\cos x = \frac{e^{ix} + e^{-ix}}{2}, \qquad \cosh x = \frac{e^x + e^{-x}}{2}$$

By direct substitution, we see immediately that

$$\sin ix = i \sinh x, \qquad \sinh ix = i \sin x$$

$$\cos ix = \cosh x, \qquad \cosh ix = \cos x \qquad (13.5\text{-}2)$$

Therefore, the analytic representation of the Chebyshev polynomials over all x values is

$$T_n(x) = \begin{cases} \cos[n \arccos x], & |x| \le 1 \\ \cosh[n \operatorname{arccosh} x], & |x| \ge 1 \end{cases} \qquad (13.5\text{-}3)$$

and when $|x| > 1$, we expect a rapid growth in $T_n(x)$ that will make the transition band from "pass" to "stop" fairly narrow.

Returning to the transfer function, we see that at the edge of the passband, $w = w_p$, we have $T_n(1) = 1$, and the condition at the edge of the passband is satisfied.

The condition at the edge of the stopband where $w = w_s$, is

$$\frac{1}{1 + \epsilon^2 T_n^2(w_s/w_p)} = \frac{1}{A^2}$$

or

$$\left| T_n\left(\frac{w_s}{w_p}\right) \right| = \frac{\sqrt{A^2 - 1}}{\epsilon}$$

Hence

$$\cosh\left(n \, \text{arccosh} \, \frac{w_s}{w_p}\right) = \frac{\sqrt{A^2 - 1}}{\epsilon}$$

Take the arccosh of both sides and divide by arccosh(w_s/w_p)

$$n = \frac{\text{arccosh}[\sqrt{A^2 - 1}/\epsilon]}{\text{arccosh}(w_s/w_p)} \qquad (13.5\text{-}4)$$

and we have the *design parameter n*.

There is an alternate representation of the arccosh x. If

$$y = \text{arccosh} \, x$$

then

$$x = \cosh y = \frac{e^y + e^{-y}}{2}$$

This result can be written as a quadratic in e^y

$$(e^y)^2 - 2x(e^y) + 1 = 0$$

whose solution is

$$e^y = x + \sqrt{x^2 - 1}$$

where we have used the + sign for $|x| > 1$. Thus taking logs

$$y = \text{arccosh} \, x = \ln[x + \sqrt{x^2 - 1}] \qquad (13.5\text{-}5)$$

As with the Butterworth filters, to obtain the transfer function $\tilde{H}(f)$ from $\tilde{H}(f)\tilde{H}(-f)$, we must find zeros of the denominator

$$1 + \epsilon^2 T_n^2\left(\frac{w}{w_p}\right) = 0$$

or

$$T_n\left(\frac{w}{w_p}\right) = \pm \frac{i}{\epsilon}$$

Let $(w/w_p) = x$. Then we have

$$\cos[n \arccos x] = \pm \frac{i}{\epsilon}$$

$$n \arccos x = \arccos\left(\pm \frac{i}{\epsilon}\right)$$

To find the value of an arccos of an imaginary number, set

$$\arccos\left(\pm \frac{i}{\epsilon}\right) = \alpha + i\beta$$

Take the cos of both sides and expand using (13.5-2)

$$\pm \frac{i}{\epsilon} = \cos(\alpha + i\beta)$$

$$\pm \frac{i}{\epsilon} = \cos \alpha \cosh \beta - i \sin \alpha \sinh \beta$$

Equating real and imaginary parts gives the two equations

$$\cos \alpha \cosh \beta = 0$$

$$\sin \alpha \sinh \beta = \mp \frac{1}{\epsilon}$$

From the first of these two equations, since $\cosh \beta \neq 0$ (real β),

$$\cos \alpha = 0$$

and therefore

$$\alpha = \frac{\pi}{2} + k\pi$$

The second equation becomes

$$\sin\left(\frac{\pi}{2} + k\pi\right) \sinh \beta = \mp\frac{1}{\epsilon}$$

$$(-1)^k \sinh \beta = \mp\frac{1}{\epsilon}$$

$$\sinh \beta = \pm\frac{(-1)^{k+1}}{\epsilon}$$

This result determines a $\beta_0 > 0$ and a $\beta_1 < 0$. Thus we have for the desired zeros

$$\arccos x = \frac{1}{n}[\alpha + i\beta_m]$$

$$= \frac{1}{n}\left[\frac{\pi}{2} + k\pi + i\beta_m\right] \qquad (m = 0, 1) \qquad (13.5\text{-}6)$$

$$x = \cos\left[\frac{\pi}{2n} + \frac{k\pi}{n} + \frac{i\beta_m}{n}\right]$$

$$= \cos\frac{\pi(2k + 1)}{2n}\cosh\frac{\beta_m}{n} - i \sin\frac{\pi(2k + 1)}{2n}\sinh\frac{\beta_m}{n}$$

and we have our zeros. There are two sequences: $m = 0, 1$ and $k = 0, 1,$ . . . , $2n - 1$. But only half actually appear (since $\beta_1 = -\beta_0$ and $\sin[\pi(2k + 1)/2n]$ has opposite signs as k comes down from $2n$ or goes up from zero). Thus the search can be confined to a single m, say $m = 1$.

The next step is to select the zeros in the upper half-plane. With our choice of m, the second term has a positive sign, and we confine the k values to 0, 1, . . . , $(n - 1)$ so as to make zeros fall in the upper half-plane.

The zeros must be paired off, from symmetric positions about the middle, and the corresponding quadratic terms formed. On finally returning to the z variable, we find that we have a product of second-order terms, numerator and denominator, plus one possible first-order term. Again we identify the coefficients with the corresponding second-order digital filter and have the cascade of filters to use in obtaining the desired, stable transfer function. This is the same computer routine as the Butterworth design—only the positions of the zeros are different.

13.6 CHEBYSHEV FILTERS, TYPE 2

The type 2 Chebyshev filters have the ripples in the stopband and are smooth in the passband (Figure 13.6-1). This time we start with the function (again in the w variable)

$$\tilde{H}(f)\tilde{H}(-f) = \cfrac{1}{1 + \epsilon^2 \left[\cfrac{T_n(w_s/w_p)}{T_n(w_s/w)}\right]^2}$$

At $w = w_p$ the condition

$$\tilde{H}(f)\tilde{H}(-f) = \frac{1}{1 + \epsilon^2}$$

is automatically satisfied. At the edge of the stopband we have $w = w_s$, and

$$\frac{1}{A^2} = \frac{1}{1 + \epsilon^2 T_n^2(w_s/w_p)}$$

Take the reciprocal of both sides:

$$A^2 = 1 + \epsilon^2 T_n^2\left(\frac{w_s}{w_p}\right)$$

Finally, we solve for n, the design parameter,

$$n = \frac{\text{arccosh}[\sqrt{A^2 - 1}/\epsilon]}{\text{arccosh}[w_s/w_p]}$$

which is the same formula as for type 1 Chebyshev filters, (13.5-4).

We next find the zeros of the denominator, take those in the upper half-plane, pair them off properly, select the second-order sections, go back to the z variable, and identify the corresponding coefficients of the second-order filters (with possibly an additional first-order filter). Again the cascade of the filters gives the transfer function sought. The details are as messy as those of the last section and are hardly more enlightening when displayed in full. Thus we omit them here. Thus all three design methods differ only in the placement of the zeros.

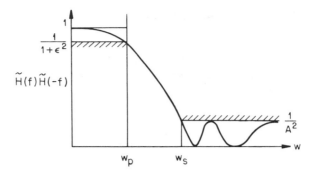

FIGURE 13.6-1 CHEBYSHEV TYPE 2 FILTER

13.7 ELLIPTIC FILTERS

Elliptic filters have ripples in both the passband and stopband, but their design, as the name implies, depends on the theory of elliptic functions and hardly belongs in an elementary course. So, regretfully, we omit them with the observation that if the reader wants such a filter there are advanced books in which they are described in some detail.

It should be evident, however, that by selecting points in both the pass- and stopbands (at roughly Chebyshev spacing), and then finding the corresponding rational function to fit this data will result in a filter that approaches equal ripple in the passband and stopband. A few iterations (see Sections 9.9 and 13.8) of placing the zeros should result in a suitable design for that particular choice of the number of zeros in the numerator and denominator. It may be worthwhile to try a few different choices if the filter is critical for further work. This design method of successive approximations is very simple and useful.

13.8 LEVELING AN ERROR CURVE

In applying a Chebyshev minimax criterion to a problem, we often find that we have an approximation whose error curve is close to, but not exactly, equal ripple. In such situations, if we want an exactly equal ripple error curve, we can use the error to find a new approximation, thereby using a feedback loop to find a level error curve ("level" means that all the local extreme errors are of the same size, not that all the errors are the same). Various methods are used, depending on personal preference. Some believe that it is merely a matter of the process of optimization, a question not of the result but of how to get there. We take the view that machine time for the average person is not important (although it may be for the specialist filter designer), so almost any plausible optimization method is useful. Library optimizers tend to have trouble with the Chebyshev criterion, since the surface of the quality-of-fit function is apt to have corners.

One of the most popular methods is to compute the position and amounts of each extreme value and then perturb each parameter a small amount, one at a time, noting the change at each extreme value in the error curve due to the perturbation. Next, a set of simultaneous linear equations is set up to make the errors equal sized, and the equations are solved for the appropriate changes in the parameters. By repeating this process a number of times, we can get arbitrarily close to the equal-level solution. See Remez algorithms in the literature.

Another method is to note the approximate locations of the zeros of the error curve, moving the zeros together to reduce the larger extremes in the error curve and moving them apart so as to increase the smaller extremes. Again, a few iterations will usually level the curve sufficiently for all practical purposes.

It should be noted, however, that almost any good optimization program will be capable of leveling the curve, *provided* the proper criteria are fed into it. For problems in which the parameters enter linearly, the approach to the best solution is generally both rapid and reliable; but when the parameters enter nonlinearly, there is no guarantee that the general case has a unique minimum; instead there may be several local minima of the maximum error. All that can be done is to find a good, plausible starting value for the first trial and hope that it is in the right valley. Using various starting functions will give an idea of the size of the local valley, for if the functions all come down to the same minimum, they are in the same valley (probably).

13.9 A CHEBYSHEV IDENTITY

The preceding completes the classical approach through the transfer function, and we now go to the direct approach of filter design through the difference-equation representation of the filter.

To carry out the design of an integrator in Section 13.10, we will need the identity

$$e^{izs} = J_0(z) + 2 \sum_{n=1}^{\infty} i^n J_n(z) T_n(s) \tag{13.9-1}$$

where the coefficients $J_n(z)$ in the expansion in Chebyshev polynomials are Bessel functions. Since this identity is not in the literature, we shall derive it along with a few properties of the Bessel functions.

We begin with the definition of the Bessel functions by means of the *generating function*

$$e^{(z/2)(t - 1/t)} = \sum_{n=-\infty}^{\infty} J_n(z)t^n \tag{13.9-2}$$

The expansion in powers of t has the Bessel functions as the coefficients. The substitution $t = -1/t'$ leaves the left side unchanged; and since the expansion is unique, it follows that

$$J_n(z) = (-1)^n J_{-n}(z) \tag{13.9-3}$$

or, as we need it,

$$(i)^{-n} J_{-n}(z) = i^n J_n(z) \tag{13.9-4}$$

For $z = 0$ we have, from the generating function, (13.9-2),

$$e^0 = 1 = \sum_{n=-\infty}^{\infty} J_n(0)t^n$$

and conclude that

$$J_0(0) = 1, \qquad J_n(0) = 0 \qquad (n \neq 0)$$

From the form

$$e^{(z/2)(t - 1/t)} = e^{z(t/2)(1 - 1/t^2)}$$

and expanding the exponential on the right it is easy to see that the Bessel functions are even or odd, depending on whether the index n is even or odd.

Expanding the exponential of the generating function (13.9-2),

$$e^{z/2(t - 1/t)} = \sum_{n=0}^{\infty} \frac{[(z/2)(t - 1/t)]^n}{n!}$$

and arranging in powers of t, we see that

$$J_n(z) = \left(\frac{z}{2}\right)^n \sum_{k=0}^{\infty} \frac{(-z^2/4)^k}{k!\,(k + n)!}$$

$$= \left(\frac{z}{2}\right)^n \left[\frac{1}{n!} - \frac{z^2/4}{1!\,(n + 1)!} + \frac{z^4/16}{2!\,(k + 2)!} - \cdots\right]$$

Set

$$t = ie^{i\theta}$$

Then the exponent of (13.9-2) is

$$z\left[\frac{t - 1/t}{2}\right] = zi\left[\frac{e^{i\theta} + e^{-i\theta}}{2}\right] = iz \cos\theta$$

and (13.9-2) becomes, when you combine positive and negative exponent terms (13.9-4)

$$e^{iz \cos\theta} = \sum_{n=-\infty}^{\infty} i^n J_n(z) e^{in\theta}$$

$$= J_0(z) + 2 \sum_{n=1}^{\infty} i^n J_n(z) \cos n\theta$$

Finally, we set $\cos\theta = s$ and we get (13.9-1), which is our fundamental identity in both s and z.

$$e^{izs} = J_0(z) + 2 \sum_{n=1}^{\infty} i^n J_n(z) T_n(s)$$

13.10   AN EXAMPLE OF THE DESIGN OF AN INTEGRATOR

Here we apply the identity in (13.9-1) to a specific example, Tick's integration formula of Section 3.4. It should be noted that Tick found the formula by computing the transfer function (the frequency response) of a family of formulas, searching until he found the one that he wanted. This method (given in Section 13.8) should not be regarded as being "too simple to use." We are merely using this problem to illustrate a design method.

The function that Tick wanted was the integral $y_n(t)$ of the measured data $y'_n(t)$. He imposed two requirements on the general form of a recursive filter

$$y_{n+1} = y_{n-1} + ay'_{n+1} + by'_n + ay'_{n-1} \qquad (13.10\text{-}1)$$

Clearly, the y'_n are the usual u_n, the integrand values. First, he required that $y'(t) = c$ be integrated exactly. This means

$$2 = 2a + b \qquad (13.10\text{-}2)$$

Second, he wanted the error curve at any frequency to be Chebyshev in the lower half of the Nyquist interval. His symmetric form meant that he had no phase errors in this recursive filter, (judged at the midpoint)

Suppose, as usual, that we have

$$y(t) = e^{i\omega t} = e^{2\pi i f t}$$

and are sampling at unit spacing. To get an expansion in Chebyshev polynomials, we use the fundamental identity (13.9-1). To use it we need to identify the variables z and s in

$$e^{2\pi i f t} = e^{izs}$$

Let the Chebyshev condition apply to the fraction λ of the Nyquist interval (we will put $\lambda = \frac{1}{2}$, finally, to get Tick's formula). Since $-1 \le s \le 1$ corresponds to $-\lambda/2 \le f \le \lambda/2$, we must have

$$s = \frac{2f}{\lambda}$$

$$z = \pi \lambda t$$

and therefore

$$e^{2\pi i f t} = J_0(\pi \lambda t) + 2 \sum_{k=1}^{\infty} i^k J_k(\pi \lambda t) T_k\left(\frac{2f}{\lambda}\right) \qquad (13.10\text{-}3)$$

Substituting it into the integration formula and assuming we are around the origin ($k = 0$), we obtain

$$J_0(\pi \lambda) + 2 \sum_{k=1}^{\infty} i^k J_k(\pi \lambda) T_k(s) = J_0(-\pi \lambda) + 2 \sum_{k=1}^{\infty} i^k J_k(-\pi \lambda) T_k(s)$$

$$+ a\pi\lambda \left[J_0'(\pi \lambda) + 2 \sum_{k=1}^{\infty} i^k J_k'(\pi \lambda) T_k(s) \right]$$

$$+ b\pi\lambda \left[J_0'(0) + 2 \sum_{k=1}^{\infty} i^k J_k'(0) T_k(s) \right]$$

$$+ a\pi\lambda \left[J_0'(-\pi \lambda) + 2 \sum_{k=1}^{\infty} i^k J_k'(-\pi \lambda) T_k(s) \right]$$

Because of the oddness and evenness of the Bessel functions, the coefficients of $T_{2k}(s)$ all vanish, leaving odd index k terms.

The coefficient of $T_1(s)$ will be zero if (dropping a $2i$ factor)

$$J_1(\pi\lambda) - J_1(-\pi\lambda) + \pi\lambda[aJ'_1(\pi\lambda) + bJ'_1(0) + aJ'_1(-\pi\lambda)]$$

or

$$2\pi\lambda aJ'_1(\pi\lambda) + \pi\lambda bJ'_1(0) = 2J_1(\pi\lambda)$$

Using $J'_1(0) = \frac{1}{2}$ together with the earlier condition (13.10-2)

$$2a + b = 2$$

gives

$$a = \frac{2J_1(\pi\lambda) - \pi\lambda}{\pi\lambda[2J'_1(\pi\lambda) - 1]} = \frac{2[J_1(\pi\lambda)/(\pi\lambda)] - 1}{2J'_1(\pi\lambda) - 1}$$

Table 13.10-1 is easily computed.

Clearly, $\lambda = 0$ is Simpson's formula. This table provides a convenient set of integration formulas whose maximum error in the lowest λth fraction of the Nyquist interval is minimal.

The result at $\lambda = \frac{1}{2}$ is very close to Tick's result ($a = 0.3584$) and differs because our error term looks like $T_3(s)$, *but* it includes higher terms $T_5(s)$, and so on, in the Chebyshev expansion of the error of fit of the two sides, whereas his was carefully leveled by repeated computation of the error curve. *At* $\lambda = 1$ the numbers in the table are about twice the Hamming window coefficients, 0.23, 0.54, 0.23.

This discussion illustrates one method of designing integrators. Many others are readily devised. Clearly, recursive filters are necessary to "remember" the initial conditions in problems like integration.

TABLE 13.10-1

λ	a	b
0.0	0.33333	1.33333
0.1	0.33425	1.33150
0.2	0.33703	1.32594
0.3	0.34177	1.31647
0.4	0.34862	1.30276
0.5	0.35785	1.28431
0.6	0.36979	1.26043
0.7	0.38493	1.23014
0.8	0.40394	1.19213
0.9	0.42771	1.14457
1.0	0.45752	1.08496

Exercises

13.10-1   Examine the family of formulas

$$y_{n+2} = y_{n-2} + ax_{n+2} + bx_{n+1} + cx_n + bx_{n-1} + ax_{n-2}$$

13.10-2   Examine the family of formulas

$$y_{n+1} = y_{n-1} + ax_{n+2} + bx_{n+1} + cx_n + bx_{n-1} + ax_{n-2}$$

13.11   PHASE-FREE RECURSIVE FILTERS

Although in general a recursive filter will have a phase angle that is not always zero, the preceding example shows, along with the trapezoid and Simpson's formulas, that some recursive filters need not have phase errors of one frequency relative to another (evaluated at the middle of the range). Evidently, even symmetry in the coefficients of the u values and odd symmetry in the coefficients of the y values (about the middle) will lead to cosine terms in the numerator and sine terms in the denominator with a spare i value, which is just what is needed to approximate an integrator.

If one is willing to admit a characteristic root on the edge of the region of stability (and we have done this), then the evenness (or oddness) of both numerator and denominator can be achieved, resulting in no phase for smoothing filters. The symmetry you need for the phase uses up about half the parameters you have available in the design, which is about what you would expect.

13.12   THE TRANSIENT

An often neglected problem when using recursive digital filters is the question of how long it takes for the transient to fade out (effectively). The recursive filter requires both an input stream of u_n and a set of initial y_n to start up. Usually these initial y_n are not known, and people have often decided to set them equal to zero (as if that were not a definite choice). These are the initial conditions and, as discussed in Section 12.3, in the mathematical solution of the difference equation they give rise to the *transient*.

How fast does the transient die out? That depends on how close one or more of the characteristic roots $|\rho_i|$ are to 1 in size. (Stability requires that none be larger than 1.) For an integrator we have, of course, at least one $\rho_i = 1$, but for smoothers generally the roots are less than 1 (otherwise the smoothing would not be "local" but would have effects that arose from very remote input values). If you look at the Butterworth and Chebyshev designs of smoothing filters, you see that the higher-degree polynomial in the denominator you use the closer a pair of complex roots are to the x axis in the z plane, and thus the closer a pair of $| \rho_i |$ are to 1. Thus there is a strong tendency for "the better the filter, the longer it takes for the transient to die out."

The problem is to figure out how to pick the initial y_n values so that effectively the transient (the solution of the homogeneous equation) will have all the $C_i = 0$ (or close to it). If we could do this, we would have the solution that depends solely on the input data u_n and not on the arbitrary start-up conditions. The fact is that we have no idea of how to do this in most cases. There have been proposals to run the filter backward for some time to try to get some starting y_n data that are consistent with the input u_n; but running a filter backward effectively takes the reciprocals of the ρ_i and makes any roots that are less than 1 into roots that are greater than 1, generally a highly unstable filter! Hence this does not seem like a very useful idea in practice.

If you have a comparatively short run of data to analyze, you have to be careful about using a recursive digital filter; you may run out of data before you have gotten rid of the spurious initial conditions! There is no simple answer to this difficulty. As a result, recursive digital filters tend to be used in situations where there is a long run of more or less homogeneous (stationary in statistical jargon) data available.

A simple experimental approach to the question of how long it takes to have the transient die out is to run the same data through the filter using a variety of different starting y_n values and all the input $u_n = 0$. Where the output tend to agree and approach 0, you are where the transient (which is the difference between any two solutions of the complete equation), is small. This is a very useful way of checking any theoretical deductions you may have made based on the size of the $| \rho_i |$.

14

Miscellaneous

14.1 TYPES OF FILTER DESIGN

Filtering problems can be conveniently divided into two extreme types. At one end are those that occur in telephony, radio, and television systems. Here the design criteria are mainly determined by the systems design, and the filter designer rarely sees the filter in action. He thinks not of the particular set of data to be processed but rather of the ensemble of signals, all having certain properties in common (primarily bandwidth restrictions).

At the other extreme is the person facing a single, unique set of data to be understood. For instance, suppose that the information consists of the number of bank failures per year in the United States for each of the past 50 years. What is wanted from the data? Understanding of the mechanisms causing the failures, of course! To do this the person analyzing the data begins with the attitude that bank failures are a function of time. Further thought may well suggest that the proper (perhaps not) thing to think about is the probability $p(t)$ of failure of a single bank as a function of time. If so, then the number of failures should be divided by the total number of banks in existence at time t. Next there comes the thought that the data is merely one realization of the mechanism of bank failures that is being sought.

Then comes the realization that you have not $p(t)$ but rather grouped data by years: thus the imagined $p(t)$ has been passed through a rectangular window of length one year, and is then sampled yearly to produce the observations on hand. This is a rectangular convolution window and corresponds to the multiplying window $\sin x/x$ in the frequency domain.

Next comes the fact that only a run of 50 years is available; this time it is a rectangular multiplying window on the data, and implies a convolving sin *x/x* in the frequency domain. Both cases have been carefully discussed in the text so that the effect on the spectrum is readily understood.

Often the data is gathered and analyzed in order to make extrapolations into the future. Polynomials are frequently used for such purposes in statistical texts, but the reader should be aware of the fact that once past the given data a polynomial will rush off to either plus or minus infinity! Thus the prediction made will tend to be either extremely pessimistic or extremely optimistic and hardly a carefully considered prediction. If, instead of blindly using the tools supplied by the textbooks, an attempt is made to understand the underlying mechanisms, a more reasonable prediction is likely to result.

The method of spectral analysis is a powerful tool for understanding linear, or almost linear, systems, but it is not suitable for the exploration of highly nonlinear systems. If the spectral approach is taken, the preceding remarks about the bank data may, in some suitable form, be relevant. In any case, the warnings in Chapter 10 about what the fast Fourier transform actually computes should be considered before any interpretation is attempted.

At this point we must stop our remarks, because we are exceeding the limits of an introduction to digital filters and are entering deeply into the field of data analysis.

Most cases fall between the two extremes cited: (1) practically infinite runs of data whose characteristics are greatly determined by the systems design, and (2) single runs of data, usually very short, and not practically extended in the near future. An alternate classification starts again with (1) the systems constraints (the properties needed for the filters to be designed), goes on to (2) where the initial data must be gathered and analyzed before the system can be designed, and (3) passes on to the attempt to find the system (as in the case of the bank failures) without going to the transfer function at all. This method of finding the system from the data will be examined briefly in Section 14.4. In this case there are two subcases that may sound the same but can be quite different: (a) you want to find the coefficients of the system, and (b) you merely want to get a system that does a good job of predicting the future. These can be quite different; you may not be able to determine the coefficients very accurately and yet be able to predict quite well, and, conversely, you may get the coefficients accurately and still not be able to predict accurately.

14.2 FINITE ARITHMETIC EFFECTS

We have discussed the computations to be done by the filter as if the arithmetic would be handled accurately and without severe roundoff. In practice, there are times when the finite word length of the numbers being used can have severe effects.

The beginner is apt to feel that although chance roundoff could occasionally affect the arithmetic, if the data is at all numerous, then he would be protected by the central limit theorems of probability, and only a very bad break would do him harm. He overlooks the fact, however, that when a coefficient of a filter is rounded off, this same roundoff occurs *every* time the coefficient is used! It is for this reason, among others, that we have recommended that the *final* filter design have its transfer function computed and drawn to see if anything peculiar has happened, to see if, because of roundoff in the coefficients, the actual filter will, in practice, differ significantly from the proposed filter.

To illustrate the strange things that can occur, consider the very simple recursive filter

$$y_n = 0.04u_n + 0.96y_{n-1}$$

with $u_m = 100$ for all m. Suppose that we start with $y_0 = 85$. We get the table [B, p. 76]

u_k	Compound y_k	Rounded y_k
100	—	85
100	85.60	86
100	86.56	87
100	87.52	88
100	88.48	88

and see that we are "stuck" at $y_n = 88$. The mathematical solution is, of course, for n approaching infinity

$$y_n = 100$$

If we start at $y_0 = 115$, we get

u_k	Computed y_k	Rounded y_k
100	—	115
100	114.40	114
100	113.44	113
100	112.48	112
100	111.52	112

and we are "stuck" at 112. Thus there is a "dead band" of width 24 units in this recursive filter. Note that the characteristic root $\rho = 0.96$ and the "settling time" is long, so the roundoff can dominate the change per cycle.

Another effect that the reader is likely to overlook consists of limit cycles. A steady input may force a recursive filter to go through a cycle of values. Overload is also a problem. Exactly what will happen when finite arithmetic is used?

All such questions become increasingly important when the filter is to be built out of definite hardware (typically, integrated circuit chips) and used in the field (instead of software programs on general-purpose computers). Unfortunately, the chip is apt to have a rather short word length (to save money and perhaps the time needed to process the numbers). In this situation, a large gap between the simulation on a computer and the actual field processing of similar data is possible. The reader is referred to advanced books, reference [IEEE-1], and the current literature, for research is still active in this field.

There is a final effect that must be considered with regard to the actual numbers used in the computation. This effect is often called the *dynamic range* of the signal. If for part of the time the signal is very small and at other times it is very large, regardless of how slow or rapid may be the rate of change, then there is a problem of accurately representing the data unless there are many digits in the representation. The reason that floating-point numbers are so common in practical computing is that they were invented to cope with this large dynamic range; they tend to give uniform relative accuracy. In such problems they are almost essential. Fortunately, more and more microcomputer chip manufacturers are recognizing the necessity for supplying floating-point arithmetic and they are gradually becoming available.

14.3 RECURSIVE VERSUS NONRECURSIVE FILTERS

We have concentrated primarily on nonrecursive filters and have discussed recursive filters in only two chapters. When should one or the other be used? As noted, recursive filters can have a very short transition band for a comparatively short span of filter. But this statement does not mean that, given a short run of data, a recursive filter is required. The transient of the filter must also be considered: how long does it take for the filter to settle down after a sudden change? Fortunately, in any particular situation the process can be simulated by a simple computation, and no great amount of theory is necessary in this very practical approach.

From a theoretical point of view, the important factor is the nearness of the zeros in the complex plane to the real frequencies (in the design-stage notation, the closeness of the zeros of the denominator to the unit circle in the z plane). If the transient of the filter takes a long time to settle down, what about properly windowing the input? About the same recommendations can be made as for computing the spectrum of data. By removing the mean (and possibly trends) and by using weights that are a cosine smoothing window, the transient can be somewhat minimized. The actual width of the cosine window is determined by the settling time of the filter (which, as noted, can be found experimentally by running a step function into the filter and simply looking at the output).

Because of these problems, as well as others like phase shifts, recursive filters tend to be used in systems where there are very long runs of data; nonrecursive filters, which are simpler to understand, design, and use (for example, no instabilities to worry about) are likely to be used in most data-processing situations where machine time is not a serious problem. Recursive filters also tend to have much less delay and are therefore used in real-time signal processing, such as telephony. Only occasionally is a recursive filter favored on the grounds that it uses less arithmetic for a sharp transition band. However, it is worth noting that both types of filters have about the same flexibility to meet various conditions when each has the same number of adjustable coefficients. It is only in one department, sharp transition bands, that the recursive filter regularly shines. Of course, special conditions can be found in which any given filter is best; we need only select its transfer function and make this the design criterion to have it win the contest.

14.4 DIRECT MODELING

Given the input u_n and the output y_n of a system of known structure, but with unknown coefficients, a model is often directly determined. Suppose, for example, the system is thought to have the structure of a recursive filter with given M and N, the number of coefficients. We simply assume the form

$$y_n = \sum_{k=1}^{M} d_k y_{n-k} + \sum_{k=0}^{N} c_k u_{n-k}$$

and the criterion of least squares. We therefore set up the difference between the two sides of the equation, square it, and write out the sum for all the cases we have data for. Do *not* supply missing zeros! Then to find the minimum least-squares fit, we simply differentiate the sum with respect to each of the unknown coefficients c_k and d_k. We get as many linear equations as there are unknowns. This system can be shown to have a nonzero determinant, so the coefficients are easily found.

If we now assume that the future is going to be like the past (that the system is stationary in statistical terms), then we have the best least-squares predictor for the future. We have a predicting filter without finding, or even talking about, the transfer function. Statisticians often do this, and neglect to examine the corresponding transfer function of the formula, which can often shed some light on the whole system. Thus, even if you do this form of direct modeling, it is often worth the extra effort to compute the corresponding transfer function.

14.5 DECIMATION

Technically, *decimation* means taking every tenth, but, in practice, it usually means taking every second one of a set. This process occurs in many problems of data processing and so a few words of warning are necessary.

It is probable that the basic reason for decimation is that in real life we do not want equally spaced values in the spectrum; instead we prefer the frequency values much more densely spaced near the zero frequency. An almost logarithmic spacing in the frequency of the spectrum is often desired. As a result, the experimenter will take a very long run of data at a fairly close spacing. The close spacing is necessary to obtain the high-frequency components of the spectrum and to avoid the aliasing that would occur if a lower sampling rate were used. The long run of data is taken in the correct belief that low frequencies will require a large number of complete periods if they are to be accurately determined.

In going from the high rate of sampling to a lower rate, we first filter the data to remove the upper half of the spectrum and then subsample by taking every other data point. This procedure avoids the aliasing that would arise from the subsampling. Frequent use of this process will reduce the volume of data to be run through the final spectrum computation to manageable amounts. The effects of the aliasing at each stage can more or less be compensated for by noting how far the filter is from ideal and multiplying the values in the upper part of the spectrum by the reciprocal of the error in the lower half of the previous spectrum.

However, note that in the early stages of decimation it is only necessary to be careful that no aliasing falls into the final band that is to be used. If other aliasing occurs, later stages of decimation will remove such frequencies.

Achieving a maximally efficient reduction is a complex problem that is not suitable for this text [CR].

14.6 TIME-VARYING FILTERS

We have discussed exclusively digital filters whose coefficients are constants. Generally, this assumes that the signal being processed is stationary. However, often the signal changes with time, and we may want a time-varying filter to process the data. The interesting case is when you do not know in advance how the statistics of the signal are going to change with time.

For example, you are trying to acquire and then predict the position of a target. In the early stages you are searching widely for the general location of the target; after it has been located, gradually you are able to track it more accurately, although you may lose it and have to start over again. The main tool for such problems is the *Kalman filter*. It is based on the least-squares approach and there is an extensive literature on it, but again it lies beyond the scope of a first text on digital filters.

Although we used the analogy of acquiring a target, a moment's thought will show that it is a very general problem that arises in many other situations. Thus you should be aware of the existence of Kalman filtering and its general use in nonstationary situations. It is sometimes called *adaptive filtering*. The performance of the filter suggests changes in its coefficients. From experience with the signal, the filter adapts itself to "better" process the signal.

14.7 REFERENCES

References are a somewhat difficult problem in an elementary book that covers a rapidly developing field of research. By the time that the reader needs the references, they may be out of date. Furthermore, the beginner is not likely to be interested in advanced references. Consequently, we have given comparatively few references in the text.

The references [IEEE-1], *Literature in Digital Signal Processing: Author and Permuted Title Index, 1975* and [IEEE-2], *Supplement to Literature in Digital Signal Processing: Author and Permuted Title Index, 1979,* are very complete as of their dates of publication, having been computer prepared and typeset. Thus the reader is referred to them as general sources of references.

Two other references, [IEEE-DSP1] and [IEEE-DSP2], are collections of published papers that were regarded as of special importance.

Presumably, further editions of the IEEE publications will occur and in time will provide the necessary up-to-date references.

The books by Oppenheim and Schafer [OS] and by Rabiner and Gold [RG] are excellent ones that start from the assumption that the analog filter theory is already known. The first is suitable for advanced undergraduates; the second is a more advanced text.

REFERENCES

[B]   BLACKMAN, R. B., *Data Smoothing and Processing.* Reading, MA: Addison-Wesley, 1965.

[B1]   BLOOMFIELD, P., *Fourier Analysis of Time Series: An Introduction.* New York: Wiley, 1976.

[BT]   BLACKMAN, R. B., and J. W. TUKEY, *The Measurement of Power Spectra.* New York: Dover, 1958.

[Br]   BRIGHAM, E. O., *The Fast Fourier Transform*. Englewood Cliffs, NJ: Prentice-Hall, 1974.

[CR]   CROCHIERE, R. E., and L. R. RABINER, "Optimum FIR Filter Implementations for Decimation, Interpolation, and Narrow-Band Filtering," *IEEE Trans. Acoustics, Speech, and Signal Processing*, pp. 444–56, October 1975.

[H]   HAMMING, R. W., *Numerical Methods for Scientists and Engineers*. New York: Dover Publications, 1986.

[IEEE-1]   *Literature in Digital Signal Processing: Author and Permuted Title Index*. Revised and expanded edition, edited by H. D. Helms, J. F. Kaiser, and L. R. Rabiner. New York: IEEE Press, 1975.

[IEEE-2]   *Supplement to Literature in Digital Signal Processing: Author and Permuted Title Index*. Supplement to the revised and expanded edition, edited by J. F. Kaiser and H. D. Helms. New York: Special publication of IEEE (JH4686–2 ASSP), 1979.

[IEEE-DSP1] *Digital Signal Processing*, edited by L. R. Rabiner and C. M. Rader. New York: IEEE Press, 1972.

[IEEE-DSP2] *Digital Signal Processing II*, edited by the Digital Signal Processing Comm. New York: IEEE Press, 1975.

[KS3]   KENDALL, M. G., and A. STUART, *The Advanced Theory of Statistics*, vol. 3. New York: Hafner, 1968.

[L]   LANCZOS, C., *Applied Analysis*. Englewood Cliffs, NJ: Prentice-Hall, 1956.

[Li]   LIGHTHILL, M. J., *Fourier Analysis and Generalized Functions*. New York: Cambridge University Press, 1960.

[P]   PAPOULIS, A., *The Fourier Integral and Its Applications*. New York: McGraw-Hill, 1962.

[T]   TUKEY, J. W., *Exploratory Data Analysis*. Reading, MA: Addison-Wesley, 1976.

Index